Hüseyin Karadeniz

Numerical Modeling of Stagnation Flows over Porous Catalytic Surfaces

Numerical Modeling of Stagnation Flows over Porous Catalytic Surfaces

by

Hüseyin Karadeniz

KIT Scientific Publishing

Dissertation, Karlsruher Institut für Technologie (KIT)
Fakultät für Maschinenbau, 2015

Tag der mündlichen Prüfung: 6. März 2015
Referenten: Prof. Dr. Ulrich Maas, Prof. Dr. Olaf Deutschmann

Impressum

KIT Scientific Publishing

Karlsruher Institut für Technologie (KIT)
KIT Scientific Publishing
Straße am Forum 2
D-76131 Karlsruhe

www.ksp.kit.edu

Print on Demand 2016

ISBN 978-3-7315-0434-4
DOI 10.5445/KSP/1000049629

Arriving at one goal is the starting point to another...

John Dewey

Acknowledgements

Firstly, I would like to express my deep gratitude to my supervisor Prof. Dr. Olaf Deutschmann for giving me the opportunity to perform my PhD study in his group. His major scientific mentor and giving me the freedom to continue my research have become the key factors for this thesis to reach to the end. I would like to deeply thank Dr. Steffen Tischer for his support and guidance during my research as well.

I deeply appreciate Prof. Dr. Ulrich Maas for accepting me as a PhD Student at mechanical engineering department.

I would like to thank Prof. Robert J. Kee (Colorado School of Mines, Golden, USA) for his very helpful guidance and scientific mentor. I also thank Dr. Huayang Zhu (Colorado School of Mines,Golden, USA) for his help regarding the dusty-gas model implementation.

I gratefully acknowledge Dr. Canan Karakaya for her explanations on stagnation-flow reactor experiments and reaction mechanisms. I thank Julian Bär for discussions on stagnation-flow reactor experiments. I also thank Dr. Lea Kahle and Karla Herrera for their useful discussions regarding the reaction mechanisms. I thank Dr. Dario Livio for his useful discussions on washcoat structures.

I specially thank Dr. Matthias Hettel for his positive comments on my studies and also very useful scientific guidance.

I thank Behnam Mozafferi and Sivaram Kannepalli for their useful discussions on numerical modeling of reactive flows.

Finally, I deeply thank my father Dr. Kemal Karadeniz for his constant support during my PhD. His scientific assistance and life guidance have always given me great confidence. I also deeply thank my mother for her constant support. I thank my brother and sister deeply for their supports, as well. I also thank all my friends who have supported me during my PhD.

Abstract

A one-dimensional stagnation-flow reactor model is developed for simulating stagnation flows on a catalytic porous plate. The flow field is coupled with the porous catalyst in its one-dimensional form. Mass transfer in the washcoat is considered for two different conditions, i.e. instantaneous diffusion (infinitely fast mass transport) and finite diffusion within the porous layer. Finite diffusion inside the washcoat is accounted by three different approaches: effectiveness factor approach, reaction-diffusion equations and dusty-gas model. Energy balance equations in the washcoat are included to investigate heat transport inside the washcoat.

A new computer code, DETCHEMSTAG, is developed to execute the numerical model. The model and computer code have the advantage (over the alternatives; CHEMKIN SPIN and DIFRUN) of incorporating different models for internal diffusion in the porous catalyst layer and coupling the diffusion model with multi-step heterogeneous reaction mechanisms. The computer code also calculates the Damköhler num-ber in stagnation flows for investigating the effect of external mass transfer limitations.

Direct oxidation of carbon monoxide over a porous Rh/Al2O3 catalyst is chosen at first as an example to apply the developed models and computational tool DETCHEMSTAG. A recently established stagnation-flow reactor is used to provide the experimental data and all necessary information to quantify the characteristics of the catalyst. The effect of internal mass transfer limitations due to a thick porous layer are discussed in detail for CO oxidation.

Subsequently, external and internal mass transfer limitations in water-gas-shift and reverse-water-gas-shift reactions over the porous Rh/Al2O3 catalyst are studied in detail. Dusty-gas model simulations are used to discuss the influence of convective flow on species transport inside the washcoat. It is discussed how the catalyst properties such as its thickness, mean pore diameter, porosity and tortuosity affect internal mass transfer limitations. The effect of pressure, flow rates, and washcoat thickness on CO consumption and internal and external mass transfer limitations is investigated. In addition, optimum working conditions for a commercial WGS catalyst are investigated.

Finally, stagnation-flow reactor model is used to study the partial oxidation and steam reforming of methane over a porous Rh/Al2O3 catalyst. The effect of internal mass transfer limitations and convective flow on species transport in the washcoat in partial oxidation and steam reforming cases is investigated. The possible reaction routes (total oxidation, steam reforming, and dry reforming) inside the catalyst are discussed in detail. The influence of pressure and flow rates on syngas production is considered as well. The influence of heat transport limitations due to a thick porous layer is also discussed.

Kurzfassung

Diese Arbeit befasst sich mit der Entwicklung eines Modells, das erstmals die eindimensionale Staupunktströmung an einer porösen katalytischen Oberfläche in einem Staupunktreaktor beschreibt. Die eindimensionale Staupunktströmung wird sowohl im Strömungsfeld als auch innerhalb der porösen katalytischen Schicht über mehrstufige Oberflächenreaktions-Mechanismen und mit den molekularen Stofftransportprozessen, der Diffusion und Leitung gekoppelt.

Der Stofftransportwiderstand am Phasenübergang zwischen Washcoat und Staupunkströmung wird für zwei unterschiedliche Grenzfälle betrachtet. Der erste Fall behandelt den äußeren Stofftransport zwischen Gasphase und Washcoat als unendlich schnell, wohingegen der zweite Fall, die Diffusion innerhalb des porösen Washcoats als endlich betrachtet. Die Diffusion im Washcoat wird durch drei verschiedene Ansätze behandelt, gemäß dem Effektivitäts-Faktor Modell, den Reaktions-Diffusions-Gleichungen und dem Dusty-Gas Modell. Die Energiebilanz innerhalb des Washcoats wird berücksichtigt, um die Wärmetransportlimitierungen im Washcoat zu untersuchen.

Ein neues Computerprogramm (DETCHEM^{STAG}) wurde entwickelt, dass das o.g. Modell implementiert. Das Modell hat den Vorteil, dass verschiedene Stofftransportmodelle für die Beschreibung der Diffusion innerhalb der porösen katalytischen Schicht verwendet werden können und das Modell über mehrstufige Oberflächenreaktions-Mechanismen gekoppelt ist. Darüber hinaus bietet das Computerprogramm den Vorteil, dass externe Stofftransportlimitierungen in der Staupunktströmung über die automatische Berechnung der Damköhler-Zahl detailliert untersucht werden können.

Die direkte Oxidation von Kohlenstoffmonoxid über einem porösen Rh/Al_2O_3 Katalysator wird in dieser Arbeit als Beispiel herangezogen, um die entwickelten Modelle und die Fähigkeiten des Computerprogramms $DETCHEM^{STAG}$ anzuwenden. Eine erst kürzlich aufgebaute Labor-Staupunktreaktoranlage wird verwendet, um experimentelle, sowie Katalysator spezifische Informationen zu quantifizieren. Die Auswirkungen von internen Stofftransportlimitierungen auf aufgrund der Dicke der porösen Schicht werden zum ersten Mal in der Literatur, detailliert für die CO Oxidation beschrieben. Anschließend werden die externen und internen Stofftransportlimitierungen bei der WGS und der RWGS Reaktion über dem porösen RH/Al_2O_3 Katalysator detailliert untersucht. Simulationen unter Verwendung des Dusty-Gas-Modells wurden durchgeführt um den Einfluss der Strömung auf den Stofftransport innerhalb des Washcoats zu untersuchen. Es wird diskutiert in wieweit Katalysator Eigenschaften wie Schichtdicke, mittlerer Porendurchmesser, Porosität und Tortuosität den inneren Stofftransportwiderstand beeinflussen. Die Auswirkungen von Druck, Strömungsgeschwindigkeit und Washcoat Dicke auf den CO-Umsatz sowie den äußeren als auch den inneren Stofftransportwiderstand für einen neuen WGS Katalysator werden zum ersten Mal in dieser Detailtiefe untersucht. Zusätzlich werden die optimalen Betriebsbedingungen eines kommerziellen WGS Katalysators untersucht. Schließlich wird das Modell verwendet, um die katalytische partielle Oxidation und die Dampfreformierung von Methan über dem porösen Rh/Al_2O_3 Katalysator zu untersuchen. Der Effekt von internen Stofftransportlimitierungen und von Strömungsverhältnissen auf den Stofftransport innerhalb des Washcoats wird diskutiert. Die möglichen Reaktionspfade (Totaloxidation, Dampfreformierung, Trocken-reformierung) innerhalb des Katalysators werden detailliert betrachtet. Die Auswirkungen von Druck und Strömungsgeschwindigkeit auf die Bildungsgeschwindigkeit von Synthesegas bei der partiellen Oxidation und bei der Dampfreformierung werden ebenfalls betrachtet. Der Einfluss von Wärmetransportlimitierungen aufgrund der porösen Schichtdicke des Washcoats werden zum ersten Mal in dieser Tiefe diskutiert.

Es wird angenommen, dass die fundamentalen Erkenntnisse, die aus dieser Arbeit hervorgehen, einen wichtigen Beitrag zum Verständnis der komplexen Prozesse in sowohl reaktionsteschnischen Anwendungen als auch in grundlegenderen Fragestellungen leisten. Deshalb wird außerdem angenommen, dass die Erkenntnisse in weiteren Modellierungsarbeiten Verwendung finden.

Table of Contents

1 Introduction

Catalysts have been used and developed to produce materials and products in sufficient quantities and at a reasonable cost over centuries [1]. The term 'catalysis' was expressed as a technical concept for homogenous and heterogeneous systems in a report published by J. J. Berzelius in 1835. Berzelius wrote in his report "It is then shown that several simple and compound bodies, soluble and insoluble, have the property of exercising on other bodies and action very different from chemical affinity. The body effecting the changes does not take part in the reaction and remains unaltered through the reaction" [2]. According to this early report, a catalyst does affect the rate of reaction but it remains unchanged during the process. It changes the rate of reaction by promoting a different molecular path [3] to overcome the activation energy barrier, which eventually requires lesser energy input.

Currently, 90% of all well-established chemical manufacturing processes, as well as new energy related research fields involve the use of heterogeneously catalyzed chemical processes from micro to macro scale [4]. Producing basic industrial chemicals, reforming crude oil, hydrogen generation, electricity generation, flame stabilization and reducing hazardous pollutants are some of the main applications of these processes. Today, the interest of industry and academia lies not only in getting the desired products, but also understanding and optimizing the involved heterogeneous reactive systems [5]. This requires a detailed knowledge of the heterogeneous surface reactions and the interaction of the active surface with the surrounding reactive flow. In this case, the steps of heterogeneous surface reactions such as adsorption, surface reactions and desorption, homogeneous gas-phase reactions, and heat and mass transport in the gas-phase as well as in the solid must be analyzed together for a complete understanding. As an example, the complex physical and chemical processes for a single channel of a catalytic combustion monolith are illustrated in Fig. 1.1 [5].

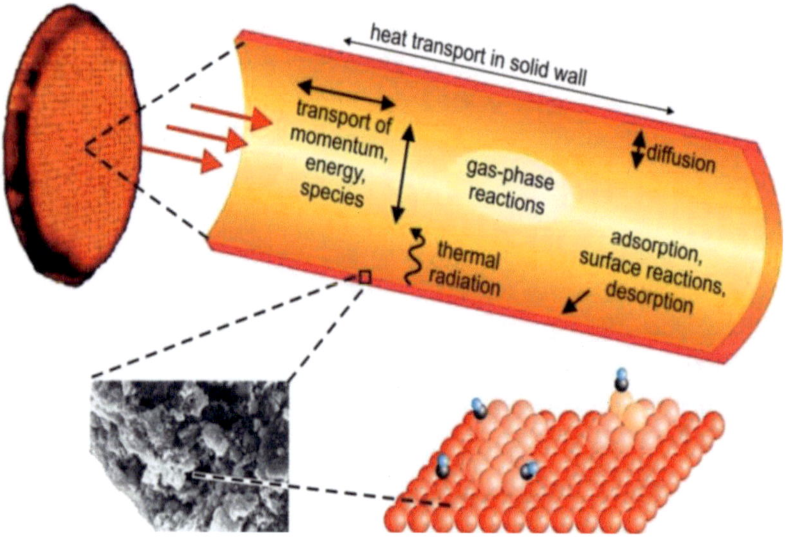

Figure 1.1: Physical and chemical processes in a single channel of a combustion monolith, the figure is taken from [5]

Computational fluid dynamics (CFD) is able to predict the behavior of chemically reactive gas-solid flows with the integration of macro and microkinetic reaction mechanisms. Macrokinetic reaction mechanisms are usually derived based on a limited range of experiments. In addition, they have very complex rate laws. In this case, microkinetic models, which are based on elementary-step reaction mechanisms, give the possibility to investigate the interactions between the reacting species on a molecular level over a wide range of temperature and pressure conditions. Therefore, they are frequently used in CFD simulations. However, solution of CFD with detailed chemistry is a challenging task due to large number of species mass conservation equations and their non-linear coupling, and the wide range of time scales caused by the complex reaction networks [5]. Experimental measurements are also needed for a complete understanding. It is also challenging to make direct experiments in the porous and narrow channels of the practical reactors. Even though turbulent flows are dominant in most technical chemical reactors with tremendous challenging for modeling and interpretation of kinetic data [6], laminar systems are favored in the kinetic studies. In this respect, the stagnation-flow reactor (SFR), which is illustrated in Fig.1.2, offers a simple configuration and a viable alternative to investigate the heterogeneously catalyzed gas-phase reactions.

In the stagnation-flow reactor (SFR) configuration, reactants are directed from the inlet manifold to the active catalytic surface through a finite gap, with a uniform flow velocity (Fig. 1.2). In general, SFR has different applications mainly based on the position of the surface, i.e., *rotating* or *non-rotating* surface. *Rotating surface* configuration is mainly used in the industry for the chemical vapor deposition (CVD) processes to grow thin solid films. A wide range of materials such as metals, alloys, silicides and nitrides have been fabricated in the industry by using the rotating disc CVD reactors [7]. *Non-rotating surface* configuration is mainly used in the SFR for measuring and modeling the gas-phase boundary layer adjacent to the zero-dimensional catalytic surface to enlighten gas-surface interactions [8, 9]. This approach is an adaptation of the frequently used approach to model counter-flow flames [10, 11]. Kinetic measurements along with the coupled model of heterogeneous chemistry with reacting flow facilitate the development of reaction mechanisms for different chemical problems such as heterogeneous catalysis [12-20], and chemical vapor deposition [21, 22]. Physical and chemical steps of heterogeneously catalyzed chemical processes, such as external and internal mass transfer limitations, and possible reaction routes in the catalyst can be investigated at a fundamental level with the integration of the developed reaction mechanisms into the appropriate numerical models. The fundamental information that is obtained through the SFR measurements and simulations can be used further for the development and optimization of practical reactors, such as monolithic reactors.

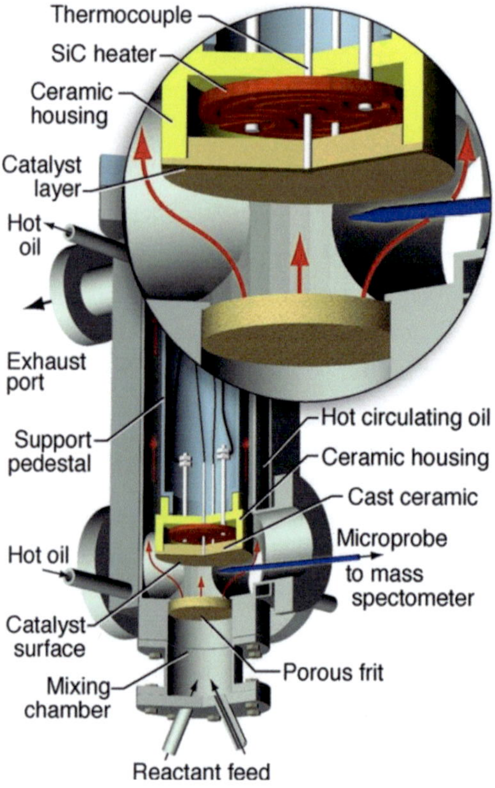

Figure 1.2: Schematic illustration of the stagnation-flow reactor, the figure is taken from [9]

The mathematical modeling of the catalytically active SFR configuration requires the incorporation of the flow equations (mass and momentum) together with the energy and species continuity equations. Chemical processes at the surface are coupled with the surrounding flow field via boundary conditions. From the modeling point of view, solving the complete Navier-Stokes equations along with energy and species continuity equations offers the most comprehensive results for the representation of the configuration, but the solution expenses also increase excessively. There are studies to develop an efficient computational solution of the partial differential equations (PDEs) regarding the three-dimensional (3D) CVD reactor-scale [23]. However, there is still a need for simplifying assumptions for the numerical models. Santen et al. [24] studied a 3D (non-axi-symmetric) and axi-symmetric consideration of stagnation-flow CVD reactors. They concluded that the existence of non-axi-symmetric flows caused by buoyancy effects occurs out of a certain value, which is obtained from a relation between Rayleigh, Reynolds and Prandtl numbers. In other words, axi-symmetric flow configuration can be considered for a certain range. Houtman et al. [7] compared complete axi-symmetric (two-dimensional) and one-dimensional (1D) stagnation-flow reactor models. They concluded that the 1D model can be applied for a wide range of conditions. Such simplified models are valid in cases where the viscous boundary-layer thickness is smaller than the lateral extent of the problem [8]. The regions, where edge effects exist, can interrupt the similarities; however one can observe that the temperature and composition fields do not vary radially in a certain regime above the deposition surface [8]. With the mentioned accurate assumptions, 1D formulation of the configuration facilitates computational modeling and simulation of processes dealing with catalytic combustion/oxidation [7, 13, 20, 25, 26].

One important feature that is used in the aforementioned catalytic combustion/oxidation studies and the practical reactor applications is the inclusion of a porous layer that is coated on the surface. The catalyst is often distributed inside a porous layer, called washcoat, to increase the internal catalyst surface area. In this case, reactants in the bulk flow diffuse from the gas-washcoat interface through the pores and react at the active sites of the catalyst. After reaction, products diffuse from the washcoat back to the bulk flow. The finite diffusion rates of the reactants and products towards and away from the active sites may lead to a reduced overall reaction rate. At low temperatures, chemical reactions are slow, and therefore their kinetics is the rate limiting step of the process. At higher temperatures, when the rate of diffusion is slow compared to the intrinsic rate of reaction, mass transport does affect the rate of reaction, and the process becomes diffusion limited [27]. As a consequence, it becomes important to include these internal mass transfer limitations in the SFR model to accurately predict the experiments, in case of a thick catalyst layer.

The main scope of this thesis is the numerical modeling of the SFR configuration over porous catalytic surfaces. The mathematical model is based on the 1D flow assumptions with energy and species continuity equations. Chemical processes at the surface are coupled with the surrounding gas-phase via boundary conditions. So far no computer code was able to account for internal mass transfer limitations in a SFR model. Mass transfer in the washcoat is considered for two different conditions, i.e. instantaneous diffusion (infinitely fast mass transport) and finite diffusion within the porous layer. Finite diffusion inside the washcoat is accounted by three different approaches: effectiveness factor approach, reaction-diffusion equations and dusty-gas model. Energy balance equations in the washcoat are included to investigate heat transport in the washcoat. Elementary-step based models for chemical reactions are included in the model. The 1D SFR model is numerically implemented by the newly developed DETCHEM^{STAG} code. DETCHEM^{STAG} is exemplarily applied in this thesis for the investigation of CO oxidation, partial oxidation, total oxidation and steam reforming of methane, water-gas-shift (WGS) and reverse water-gas-shift (RWGS) reactions at various temperatures over Rh/Al_2O_3 catalyst. In this respect, gas-phase boundary layer thicknesses and the validity of the 1D model, the effect of internal mass transfer limitations in the washcoat, the importance of pressure gradients and heat transfer limitations in the washcoat are discussed. It is discussed how washcoat parameters such as its thickness, mean pore diameter, porosity and tortuosity affect internal mass transfer limitations. In addition, the effect of external mass transfer limitations in the gas-phase is investigated. The effect of pressure and flow rates on CO consumption in WGS reaction, and syngas production in catalytic partial oxidation (CPOX) and SR of CH_4 are considered as well. It is expected that the fundamental information that is proposed in this thesis can help to understand the complex processes in practical reactor applications and new energy related research studies. Therefore, it is also expected that these fundamental information can be used in further modeling and simulation efforts regarding the heterogeneously catalyzed chemical processes.

2 Fundamentals

In this chapter, essentials of chemically reacting flow over a porous catalytic surface are described. In order to obtain a better understanding, physico-chemical steps of heterogeneous catalytic processes are explained first. These steps are summarized below and they closely follow the explanation indicated in [3]:

1) *External mass transfer of the reactants from the bulk flow to the gas-surface interface:*

   This step is necessary for the transport of the reactants to the external surface of the catalyst. Both convective and diffusive transport must be taken into consideration. Gas-phase reactions should also be included over a certain temperature. In this step, a reactant i at a bulk concentration $c_{i,\mathrm{b}}$ move through the boundary layer thickness δ to the external surface of the catalyst where the concentration is $c_{i,s}$. If the fluid flow over the external washcoat is slow, the boundary layer across which species i must be transported becomes thick, and it takes a long time for species i to travel to the surface [3]. Therefore, mass transfer across the boundary layer becomes slow and it limits the rate of the overall reaction [3]. The impact of the external mass transfer limitations is discussed later regarding the stagnation flow simulations.

2) *Mass transport of the reactants from the gas-surface interface through the catalyst pores to the intermediate vicinity of the internal catalytic surface:*

   This internal diffusion step is necessary for reactants to be adsorbed on available active sites within the washcoat. If the concentration gradient inside the washcoat becomes large due to the internal mass transport limitations, this step determines the overall reactivity. The impact of the internal mass transfer limitations is discussed later regarding the stagnation flow simulations.

3) *Adsorption of the reactants onto the catalyst surface:*

   This step is necessary for the surface reactions. It is modeled in reaction mechanisms commonly with the *Langmuir-Hinselwood* mechanism via mean-field approximation. This model assumes that the gas-phase species M and U adsorb on the active metal sites of the catalyst, forming $M(s)$ and $U(s)$. Then the reaction proceeds only between the adsorbed species.

4) *Surface reaction on the catalyst surface:*

   In this step, the reactants react on the active sites of the catalyst to form the products.

5) *Desorption of the products from the catalyst surface:*

   In this step, the products, which are formed via surface reactions between the adsorbed species, are desorbed from the active sites.

6) *Diffusion of the products from the intermediate vicinity of the internal catalyst surface to the external catalyst surface (gas-washcoat interface):*

   In this step, the products travel from the inner washcoat to the gas-washcoat interface.

7) *Mass transfer of the product B from the gas-washcoat interface to the bulk flow:*

   This step is necessary for the products to be transported from the external catalyst surface through the boundary layer thickness in the gas-phase.

These processes are also depicted in Figure 2.1.

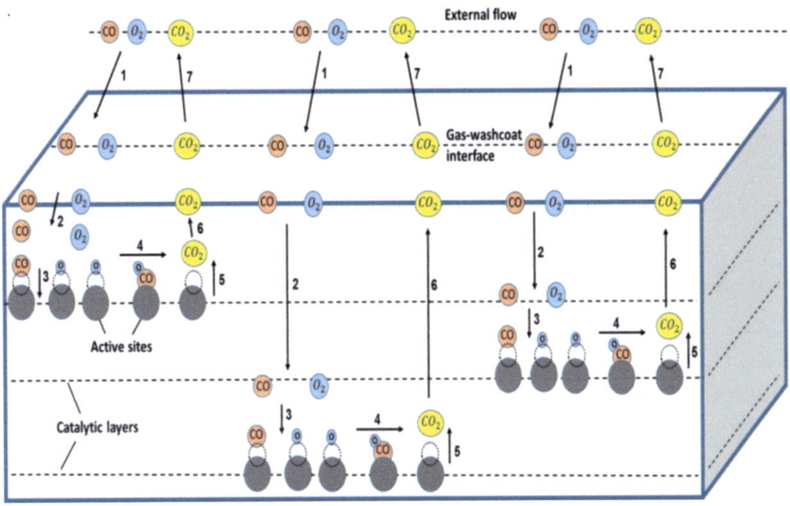

Figure 2.1: Physico-chemical Steps of Heterogeneous Catalytic Processes

Mathematical modeling of aforementioned physico-chemical steps requires considering general conservation equations of a chemically reacting flow, transport equations and chemical reactions in the gas-phase and in the porous catalyst. In addition, catalytically active surface must be closely coupled with the surrounding flow field [28]. In this respect, the following sequences are followed in this chapter for brief fundamental explanations, respectively:

- Section 2.1: chemically reactive flows
- Section 2.2: catalytic surfaces
- Section 2.3: homogeneous gas-phase and heterogeneous surface reactions
- Section 2.4: mass transport in the porous catalytic surface
- Section 2.5: heat transport in the porous catalytic surface
- Section 2.6: coupling porous catalytic surface with the surrounding reactive flow

2.1 Chemically Reacting Flows

Chemically reacting flows can be completely described if density, pressure, temperature and velocity of the mixture and concentration of each individual species are known at each point in space and time [29]. These properties can change in space and time as a result of fluid flow (convection), chemical reactions and molecular transport (conduction, diffusion and viscosity) [30]. Mathematical description and the corresponding numerical treatment of chemically reacting flows requires considering a set of conservation equations for energy, total mass, momentum and species mass fractions, which all together form the Navier-Stokes equations. Navier-Stokes equations are second-order, nonlinear and partial differential equations. Therefore, it is hard to obtain their analytical solutions. However, they can be solved numerically to predict to velocity, pressure and temperature field for a flow for known geometry and boundary conditions. Solving Navier-Stokes equations gives the density, pressure, temperature and velocity components at each point in space and time in the flow field. In the following sections of this chapter, conservation equations of chemically reacting flows are explained briefly, and they are given generally in cylindrical coordinates.

2.1.1 Conservation Laws for a System and a Control Volume

Conservation laws of fluid mechanics are mainly adapted from conservation laws of solid mechanics, which considers the change of extensive variables in unit time in *systems* [31]. Since it is more convenient to work with control volume in fluid mechanics, changes in the control volume and in the system must be related. The relation between the change of an extensive variable for a control volume and a system can be defined with Reynolds transport theorem (RTT), as it was given in [31]:

$$\frac{dB_{sys}}{dt} = \frac{dB_{cv}}{dt} - \dot{B}_{in} + \dot{B}_{out} \tag{2.1}$$

where B is any extensive property, and $b = B/m$ is the corresponding intensive property. The equation simply means that the change of variable B in the system is equal to the change of B in the control volume, and net flux of B with the mass flowing from the control surface [31].

The flow rate of b, which is flowing from a differential surface area ($\mathrm{d}A$) on the surface (Fig. 2.2), can be calculated from $\rho b\vec{\mathrm{V}} \cdot \vec{\mathrm{n}}\mathrm{d}A$, as it was given in [31]

$$\dot{B}_{net} = \dot{B}_{out} - \dot{B}_{in} = \int_{CS} \rho b\vec{V} \cdot \vec{n} dA \tag{2.2}$$

in which $\vec{\mathrm{n}}$ is the unit vector.

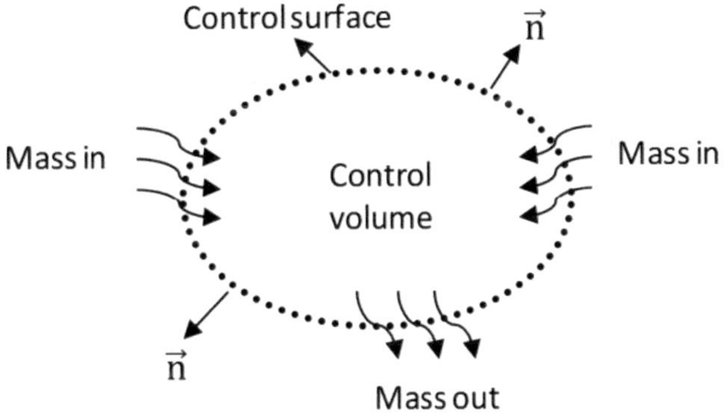

Figure 2.2: Schematic illustration of the flow rate of B to the control volume through the control surface

Since the properties in the control volume might change with the location, total amount of property B in the control volume can be considered with Eq. (2.3), as it was given in [31],

$$B_{CV} = \int_{CV} \rho b dV \tag{2.3}$$

which means that dB_{cv}/dt will be equal to $\frac{d}{dt}\int_{cv} \rho b dV$. Now if Eq. (2.2) and Eq. (2.3) are put in Eq. (2.1), general form of the RTT is obtained as [31, 32].

$$\frac{dB_{sys}}{dt} = \int_{CV} \frac{\partial}{\partial t}(\rho b) dV + \int_{CS} \rho b \vec{V} \cdot \vec{n} dA \quad . \tag{2.4}$$

In the following conservation equation explanations, RTT will be used to relate the change of an extensive variable for a system and control volume.

2.1.1.1 Conservation of Mass (Continuity Equation)

Conservation of mass can be derived from Eq. (2.4). In a *closed system*, the mass of the *system* remains constant during the process ($m_{sys} = \text{constant}$ or $dm_{sys}/dt = 0$). General form of the conservation of mass (continuity) is given then as

$$\int_{CV} \frac{\partial \rho}{\partial t} dV + \int_{CS} \rho \vec{V} \cdot \vec{n} dA = 0 \quad . \tag{2.5}$$

According to Eq. (2.5), sum of the change of mass with time in *control volume* and net mass transport from the control surface is zero. It is then possible to state Eq. (2.5) in a differential form by using the divergence theorem. Divergence theorem allows converting the divergence of the volume integral of a vector to a surface integral [32,33]. Divergence of any $\vec{G}$ vector ($\nabla \cdot \vec{G}$) is given as

$$\int_{V} \nabla \cdot \vec{G} \, dV = \int_{S} \vec{G} \cdot \vec{n} \, dA \quad . \tag{2.6}$$

Conservation of mass (continuity) can be arranged then by using the divergence theorem

$$\int_{CV} \frac{\partial \rho}{\partial t} dV + \int_{CV} \nabla \cdot (\rho \vec{V})\, dV = 0 \quad . \tag{2.7}$$

Further arrangement gives

$$\int_{CV} \left[\frac{\partial \rho}{\partial t} + \nabla \cdot (\rho \vec{V})\right] dV = 0 \quad . \tag{2.8}$$

Finally, the general form of the conservation of mass or continuity equation is obtained as [34-36]

$$\frac{\partial \rho}{\partial t} + \nabla \cdot (\rho \vec{V}) = 0 \quad . \tag{2.9}$$

Mass continuity equation can be written in cylindrical coordinates as

$$\frac{\partial \rho}{\partial t} + \frac{1}{r}\frac{\partial r \rho v_r}{\partial r} + \frac{1}{r}\frac{\partial \rho v_\theta}{\partial \theta} + \frac{\partial \rho v_z}{\partial z} = 0 \quad . \tag{2.10}$$

At steady-state formulation, the term $\partial\rho/\partial t$ vanishes in Eq. (2.10).

2.1.1.2 Conservation of Momentum

Conservation of momentum can be derived based on the Newton's second law as

$$\sum \vec{F} = \frac{d}{dt}\int_{sys} \rho \vec{V} dV \tag{2.11}$$

in which $\sum \vec{F}$ is the sum of the forces on a system *and* $\rho\vec{V}$ is the linear momentum of the system. Eq. (2.11)means that the sum of the external forces on a *system* is equal to change of momentum in unit time. Conservation of momentum can be extended to control volume by using RTT [34, 37] as

$$\sum \vec{F} = \sum \vec{F}_{body} + \sum \vec{F}_{surface} = \int_{CV} \frac{\partial}{\partial t}(\rho \vec{V}) dV + \int_{CS} \rho \vec{V}(\vec{V} \cdot \vec{n}) dA \quad . \tag{2.12}$$

Eq. (2.12) means that the sum of the external forces (body and surface forces) on a control volume is equal to sum of the change of momentum in control volume and net momentum flow rate from the control surface. Derivation of the external forces (body and surface forces) is given in the Appendix A. Using the divergence theorem and inserting external forces into Eq. (2.12), gives the differential form of the momentum conservation equation as [34,35]

$$\frac{\partial}{\partial t}(\rho\vec{V}) + \nabla\cdot(\rho\vec{V}\otimes\vec{V}) = \rho\vec{g} + \nabla\cdot T \quad . \tag{2.13}$$

in which $\vec{V}\otimes\vec{V}$ is dyadic product and T is the stress tensor. T is written in terms of pressure and velocity field as [34]

$$T = -pI + \mu\left[\nabla\vec{V} + \left(\nabla\vec{V}\right)^{T}\right] + \kappa(\nabla\cdot\vec{V})I \tag{2.14}$$

where μ is the dynamic viscosity, κ is bulk viscosity, $\nabla\vec{V}$ is the velocity-gradient tensor, $\left(\nabla\vec{V}\right)^{T}$ is its transpose and I is the unit tensor. It is referred to [33] for dyadic product and unit tensor definitions. The bulk viscosity is often taken as $\kappa = -2/3\mu$. The first term of Eq. (2.14) describes the hydrostatic part of stress tensor, the second term the viscous part and the third term the fluid dilatation part. Eq. (2.13) is valid at any point in the flow domain for compressible as well as incompressible flows. It forms along with the continuity equation the Navier-Stokes equations for nonreactive flows. The stress tensor in Eq. (2.13) has 6 unknowns, therefore it is not practical to implement. In this case, expanding the viscous stresses in terms of strain rate tensor gives the following equations for compressible flows in the cylindrical coordinates (expanding the viscous terms in terms of strain rate is explained in Appendix A) [34]:

In r-component:

$$\rho\left(\frac{\partial v_r}{\partial t}+v_r\frac{\partial v_r}{\partial r}+\frac{v_\theta}{r}\frac{\partial v_r}{\partial \theta}+v_z\frac{\partial v_r}{\partial z}-\frac{v_\theta^2}{r}\right)=-\frac{\partial p}{\partial r}+\rho g_r$$
$$+\left[\frac{\partial}{\partial r}\left(2\mu\frac{\partial v_r}{\partial r}+\kappa\nabla\cdot\vec{V}\right)+\frac{1}{r}\frac{\partial}{\partial \theta}\left(\mu\left(\frac{\partial v_\theta}{\partial r}-\frac{v_\theta}{r}+\frac{1}{r}\frac{\partial v_r}{\partial \theta}\right)\right)\right.$$
$$\left.+\frac{\partial}{\partial z}\left(\mu\left(\frac{\partial v_r}{\partial z}+\frac{\partial v_z}{\partial r}\right)\right)+\frac{2\mu}{r}\left(-\frac{1}{r}\frac{\partial v_\theta}{\partial \theta}-\frac{v_r}{r}+\frac{\partial v_r}{\partial r}\right)\right] \tag{2.15}$$

In θ-component:

$$\rho\left(\frac{\partial v_\theta}{\partial t}+v_r\frac{\partial v_\theta}{\partial r}+\frac{v_\theta}{r}\frac{\partial v_\theta}{\partial \theta}+v_z\frac{\partial v_\theta}{\partial z}+\frac{v_r v_\theta}{r}\right)=-\frac{1}{r}\frac{\partial p}{\partial \theta}+\rho g_\theta$$
$$+\left[\frac{\partial}{\partial r}\left(\mu\left(\frac{\partial v_\theta}{\partial r}-\frac{v_\theta}{r}+\frac{1}{r}\frac{\partial v_r}{\partial \theta}\right)\right)+\frac{1}{r}\frac{\partial}{\partial \theta}\left(2\mu\left(\frac{1}{r}\frac{\partial v_\theta}{\partial \theta}+\frac{v_r}{r}\right)+\kappa\nabla\cdot\vec{V}\right)\right.$$
$$\left.+\frac{\partial}{\partial z}\left(\mu\left(\frac{\partial v_\theta}{\partial z}+\frac{1}{r}\frac{\partial v_z}{\partial \theta}\right)\right)+\frac{2\mu}{r}\left(\frac{1}{r}\frac{\partial v_r}{\partial \theta}+\frac{\partial v_\theta}{\partial r}-\frac{v_\theta}{r}\right)\right] \tag{2.16}$$

In z-component:

$$\rho\left(\frac{\partial v_z}{\partial t}+v_r\frac{\partial v_z}{\partial r}+v_z\frac{\partial v_z}{\partial z}+\frac{v_\theta}{r}\frac{\partial v_z}{\partial \theta}\right)=-\frac{\partial p}{\partial z}+\rho g_z$$
$$+\left[\frac{1}{r}\frac{\partial}{\partial r}\left(r\mu\left(\frac{\partial v_r}{\partial z}+\frac{\partial v_z}{\partial r}\right)\right)+\frac{1}{r}\frac{\partial}{\partial \theta}\left(\mu\left(\frac{\partial v_\theta}{\partial z}+\frac{1}{r}\frac{\partial v_z}{\partial \theta}\right)\right)\right.$$
$$\left.+\frac{\partial}{\partial z}\left(2\mu\frac{\partial v_z}{\partial z}+\kappa\nabla\cdot\vec{V}\right)\right] \tag{2.17}$$

where $\nabla\cdot\vec{V}$ refers the fluid dilatation [34]. In case of incompressible flow, the term $\nabla\cdot\vec{V}$ vanishes. Velocity components are given here with respect to r, θ and z coordinates as $v_{r/\theta/z}$. At steady-state formulation, the terms $\partial v_r/\partial t$, $\partial v_\theta/\partial t$ and $\partial v_z/\partial t$ vanishes in Eq. (2.15), Eq. (2.16) and Eq. (2.17) respectively.

2.1.1.3 Species Conservation

In a chemically reacting flow, conservation of individual species must also be considered, because each component has its own mass density and velocity. In this case, the extensive variable is the mass of the species i (m_i), and the intensive variable is the species mass fraction of the species i ($Y_i = m_i/m$). Chemical reactions result in production or consumption of species, which can be modeled as a mass source or sink for the ith species, i.e.,

$$\left.\frac{dm_i}{dt}\right|_{sys} = \int_{CV} \dot{\omega}_i M_i \, d\mathrm{V} \tag{2.18}$$

where $\dot{\omega}_i$ and M_i are the molar production rate and molar mass of species i, respectively. By using the RTT theorem, species mass fraction for a control volume is given as [34, 38]

$$\int_{CV} \dot{\omega}_i M_i \, d\mathrm{V} = \int_{CV} \frac{\partial}{\partial t}(\rho Y_i) d\mathrm{V} + \int_{CS} \rho Y_i (\vec{V}_i \cdot \vec{n}) dA \tag{2.19}$$

Species can cross from the control surface via convection or diffusion. In this respect, second term on the right hand side of Eq. (2.19) can be rearranged as

$$\int_{CS} \rho Y_i (\vec{V}_i \cdot \vec{n}) dA = \int_{CS} \rho Y_i (\vec{V}_i \cdot \vec{n}) dA + \int_{CS} \vec{j}_i \cdot \vec{n} dA \quad . \tag{2.20}$$

where $\vec{j}_i$ is the diffusive mass-flux vector. If Eq. (2.20) is inserted into Eq. (2.19) and the divergence theorem is used, one gets

$$\frac{\partial}{\partial t} \rho Y_i + \nabla \cdot (\rho Y_i \vec{V}_i) = -\nabla \cdot \vec{j}_i + \dot{\omega}_i M_i \quad . \tag{2.21}$$

Species continuity equation can be given in the cylindrical coordinates as:

$$\rho\left(\frac{\partial Y_i}{\partial t} + v_r\frac{\partial Y_i}{\partial r} + \frac{v_\theta}{r}\frac{\partial Y_i}{\partial \theta} + v_z\frac{\partial Y_i}{\partial z}\right) = -\frac{1}{r}\frac{\partial}{\partial r}\left(r j_{i,r}\right) + \frac{1}{r}\frac{\partial j_{i,\theta}}{\partial \theta} + \frac{\partial j_{i,z}}{\partial z} + \dot{\omega}_i M_i \tag{2.22}$$

where $j_{i,r/\theta/z}$ are the diffusive mass flux of species i in the r, θ and z direction, respectively. Diffusive mass flux derivation is explained in section 2.1.2.1. At steady-state formulation, the term $\partial Y_i/\partial t$ vanishes in Eq. (2.22).

2.1.1.4 Conservation of Energy

In chemically reacting flows, thermal energy equation is used to describe and predict the fluid temperature fields. In general, thermal energy equation is derived by subtracting the mechanical energy contribution from the total energy equation. Therefore, initially total energy equation is considered.

According to the first law of thermodynamics, total energy of a *system* can change with heat transfer Q or work done on the system W. Conservation of total energy in unit time can be written for a system as

$$\frac{dE_{sys}}{dt} = \frac{d}{dt}\int_{sys} \rho\, e_t dV = \dot{Q} + \dot{W} \quad . \tag{2.23}$$

Conservation of energy can be extended to control volume by using RTT as [34, 36]

$$\dot{Q} + \dot{W} = \frac{d}{dt}\int_{CV} e_t \rho dV + \int_{CS} e_t\, \rho(\vec{V}\cdot\vec{n})dA \tag{2.24}$$

where e_t is the total specific energy. It has three contributions; internal energy, kinetic energy and potential energy of the fluid [34]. Heat can cross from the surfaces of control volume in two different ways; heat conduction via Fourier's law or energy transport via chemical species diffusion. Rate of work on the surfaces of a control volume is caused by the stress tensor. Based on these considerations, Eq. (2.24) can be arranged as

$$\int_{CS} \lambda \nabla T \cdot \vec{n} dA - \sum_{i=1}^{N_g} \int_{CS} h_i \vec{j}_i \cdot \vec{n} dA + \int_{CS} (T \cdot V) \cdot \vec{n} dA$$
$$= \frac{d}{dt} \int_{CV} e_t \rho d\mathrm{V} + \int_{CS} e_t \, \rho(\vec{V} \cdot \vec{n}) dA \tag{2.25}$$

where λ and T are the thermal conductivity and temperature of the mixture, respectively. h_i is the specific enthalpy of species i. N_g is the number of gas-phase species. Eq. (2.25) can be rearranged by using the divergence theorem as [34, 38]

$$\frac{\partial \rho e_t}{\partial t} + \nabla \cdot (\rho \, e_t \vec{V}) - \nabla \cdot (\lambda \nabla T) + \sum_{i=1}^{N_g} \nabla \cdot (h_i \vec{j}_i) - \nabla \cdot (\vec{V} \cdot T) = 0 \quad . \tag{2.26}$$

A general thermal energy equation can be obtained by subtracting mechanical energy contribution from the total energy equation [34]

$$\left(\frac{\partial \rho h}{\partial t} + \nabla \cdot (\rho h \vec{V}) \right) = \frac{\partial p}{\partial t} + \vec{V} \cdot (\nabla p) + \nabla \cdot (\lambda \nabla T) - \sum_{i=1}^{N_g} \nabla \cdot h_i \vec{j}_i + \phi \tag{2.27}$$

where ϕ is the dissipation function. Further, a perfect-gas thermal energy equation can be simplified as it was given in [34]

$$\rho c_p \left(\frac{\partial T}{\partial t} + \vec{V} \cdot (\nabla T) \right) = \frac{\partial p}{\partial t} + \vec{V} \cdot (\nabla p) + \nabla \cdot (\lambda \nabla T)$$
$$- \sum_{i=1}^{N_g} c_{p,i} \vec{j}_i \cdot \nabla T - \sum_{i=1}^{N_g} h_i \dot{\omega}_i Y_i + \phi \tag{2.28}$$

in which c_p is the specific heat capacity of the mixture. Eq. (2.28) accounts for temperature change due to mechanical compression, heat conduction, heat transport due to species diffusion, heat release due to chemical reactions and viscous dissipation. Consequently, a thermal energy equation can be given in cylindrical coordinates as

$$\rho c_p\left(\frac{\partial T}{\partial t}+v_r\frac{\partial T}{\partial r}+\frac{v_\theta}{r}\frac{\partial T}{\partial \theta}+v_z\frac{\partial T}{\partial z}\right)$$
$$=\frac{\partial p}{\partial t}+\left(v_r\frac{\partial p}{\partial r}+\frac{v_\theta}{r}\frac{\partial p}{\partial \theta}+v_z\frac{\partial p}{\partial z}\right)-\frac{1}{r}\frac{\partial}{\partial r}\left(j_{q,r}^c\right)-\frac{1}{r}\frac{\partial}{\partial \theta}\left(j_{q,\theta}^c\right)$$
$$-\frac{\partial}{\partial z}\left(j_{q,z}^c\right)-\sum_{i=1}^{N_g}c_{p,i}\left(j_{i,r}\frac{\partial T}{\partial r}+\frac{j_{i,\theta}}{r}\frac{\partial T}{\partial \theta}+j_{i,z}\frac{\partial T}{\partial z}\right)-\sum_{i=1}^{N_g}h_i\dot{\omega}_iY_i$$
$$+\phi \tag{2.29}$$

where $\mathrm{j}_{q,r/\theta/z}^{c}$ are the heat flux in the r, θ and z direction, respectively. Heat flux derivation is explained in section 2.1.2.2. At steady state formulation, the terms $\partial T/\partial t$ and $\partial p/\partial t$ vanishes in Eq. (2.29).

2.1.1.5 Ideal Gas Law

Ideal gas law relates the state variables (pressure p, density ρ, temperature T and species composition) in the conservation equation system of chemically reacting flows

$$p=\rho\frac{R}{\overline{M}}T \tag{2.30}$$

in which $\overline{M}$ is the average molar mass of the mixture and it is stated as

$$\overline{M}=\frac{1}{\sum\frac{Y_i}{M_i}} \tag{2.31}$$

where R is the universal gas constant.

2.1.2 Molecular Transport Processes

It can be seen in the conservation equations that the physical properties (mass, heat, momentum) of a fluid are transported by convective and molecular processes. Convective transport is a physical process where the physical properties of the fluid are transported from one location to another by the movement of the fluid elements. Molecular transport of the physical properties occurs due to their gradients between two neighboring gas layers in a system [30]. Their motion is a complete disorder (molecular chaos) [30].

The complexity of the molecular transport processes does not allow a purely theoretical fundamental approach. Therefore, semi-empirical concepts (based partly on experiments) are incorporated for adequately describing the molecular transport processes. In this case, transport coefficients, i.e. diffusion coefficients, thermal conductivities and viscosity coefficients, are calculated from the transport coefficients of the individual species. Mass fluxes, heat fluxes and momentum fluxes are described then based on the transport coefficients.

2.1.2.1 Diffusion

Diffusion refers to a process in which molecules of a mixture move from the regions of higher concentration into the regions of lower concentration. According to the Fick's law, diffusive mass flux is proportional to the concentration gradient [30, 39]. In the conservation equations, diffusive mass fluxes are considered in species continuity and thermal energy equations. Diffusive mass flux of a species can be given based on a mixture averaged diffusion coefficient approach as

$$\vec{j}_i^{\,d} = -\rho D_{i,M} \frac{Y_i}{X_i} \nabla X_i \tag{2.32}$$

where $D_{i,\mathrm{M}}$ and X_i are the averaged diffusion coefficient and mole fraction of the species i, respectively. $D_{i,\mathrm{M}}$ is calculated in a mixture as

$$D_{i,M} = \frac{1 - Y_i}{\sum_{j \neq i} \frac{X_j}{D_{ij}}} \tag{2.33}$$

where D_{ij} is the binary diffusion coefficient. It states the diffusion of species i in species *j*, and it is calculated from the kinetic theory of diluted gases by Chapman-Enskog correlation, as it was given in [40]

$$D_{ij} = \frac{3}{16} \frac{\sqrt{2\pi k_B^3 T^3 / m_{ij}}}{p\pi \Theta_{ij}^2 \Omega_{ij}^{(1,1)*}(T_{ij}^*)} \tag{2.34}$$

in which m_{ij}, Θ_{ij}, T_{ij}^* and $\Omega_{ij}^{(1,1)*}$ are the reduced mass, length-scale in the interaction between two molecules, reduced temperature and temperature dependence of the collision integral according to Lennard-Jones potential, respectively [40].

Diffusive mass flux due to concentration gradient is the driving force of diffusion. However, mass can also be transported due to temperature gradient between the gas layers of the mixture (*thermal diffusion* or *Soret effect*). In this case, diffusive mass flux due to *thermal diffusion* is calculated from

$$\vec{j}_i^T = \frac{D_i^T}{T} \nabla T \tag{2.35}$$

where D_i^{T} is the thermal diffusion coefficient of the species i. Thermal diffusion is important just for the light species (H and H_2) in a mixture. If both concentration gradient and thermal diffusion are taken into consideration, the diffusion flux of the species i can be given as [30, 39]

$$\hat{j}_i = \vec{j}_i^{\,d} + \vec{j}_i^{\,T} = -\rho D_{i,M} \frac{Y_i}{X_i} \nabla X_i + \frac{D_i^T}{T} \nabla T \tag{2.36}$$

Mass conservation requires that the term $\sum_{i=1}^{N_g} Y_i \, j_i$ must be equal to zero. However, this case is not always fulfilled as a result of the Fickian mixture averaged diffusion coefficient ($D_{i,M}$) [26]. Therefore, diffusion velocities are corrected using,

$$\vec{j}_i = \hat{j}_i - Y_i \sum_{k=1}^{N_g} \hat{j}_k \tag{2.37}$$

Eventually, diffusive mass flux of species i can be given in cylindrical coordinates as

$$j_{i,r} = -\rho D_{i,M} \frac{Y_i}{X_i} \frac{\partial X_i}{\partial r} + \frac{D_i^T}{T} \frac{\partial T}{\partial r} \tag{2.38}$$

$$j_{i,\theta} = -\rho D_{i,M} \frac{Y_i}{X_i} \frac{1}{r} \frac{\partial X_i}{\partial \theta} + \frac{D_i^T}{T} \frac{1}{r} \frac{\partial T}{\partial \theta} \tag{2.39}$$

$$j_{i,z} = -\rho D_{i,M} \frac{Y_i}{X_i} \frac{\partial X_i}{\partial z} + \frac{D_i^T}{T} \frac{\partial T}{\partial z} \tag{2.40}$$

2.1.2.2 Heat Transport

Molecular heat transport occurs due to heat conduction, species diffusion and Dufour effect. Heat conduction refers to a process in which heat is transported from the regions of higher temperature towards the regions of lower temperature. According to the Fourier law of heat conduction, heat flux is proportional to temperature gradient as [30, 41]

$$\vec{j}_q^{\,c} = -\lambda \nabla T \quad . \tag{2.41}$$

Thermal conductivity of the mixture is calculated from the thermal conductivity of each individual species in the mixture according to empirical law [30, 42]

$$\lambda = \frac{1}{2}\left[\sum_i X_i\lambda_i + \left(\sum_i \frac{X_i}{\lambda_i}\right)^{-1}\right] \tag{2.42}$$

where λ_i is the thermal conductivity of the species i. It is calculated from the transfer of translational, rotational and vibrational energy between molecules [43, 44]:

$$\lambda_i = \frac{\eta_i}{M_i}\left(f_{tr}c_{V,tr} + f_{rot}c_{V,rot} + f_{vib}c_{V,vib}\right) \tag{2.43}$$

where

$$f_{tr} = \frac{5}{2}\left(1 - \frac{2}{\pi}\frac{c_{V,rot}}{c_{V,tr}}\frac{A}{B}\right) \tag{2.44}$$

$$f_{rot} = \frac{\rho D_{ii}}{\eta_i}\left(1 + \frac{2}{\pi}\frac{A}{B}\right) \tag{2.45}$$

$$f_{vib} = \frac{\rho D_{ii}}{\eta_i} \tag{2.46}$$

with

$$A = \frac{5}{2} - f_{vib} \qquad and \qquad B = Z_{rot} + \frac{2}{\pi}\left(\frac{5}{3}\frac{c_{V,rot}}{R} + f_{vib}\right) \tag{2.47}$$

Z_{rot} is here a characteristic parameter and calculated proportional to the value of Z_{rot} at 298 K as

$$\frac{1}{Z_{rot}(T^*)} \propto 1 + \frac{\pi^{\frac{3}{2}}}{2}T^{*-\frac{1}{2}} + \left(\frac{\pi^2}{4} + 2\right)T^{*-1} + \pi^{\frac{3}{2}}T^{*-\frac{3}{2}} \tag{2.48}$$

Molecular heat transport due to heat conduction can be given in the cylindrical coordinates as

$$j_{q,r}^{c} = -\lambda \frac{\partial T}{\partial r} \tag{2.49}$$

$$j_{q,\theta}^{c} = -\lambda \frac{1}{r}\frac{\partial T}{\partial \theta} \tag{2.50}$$

$$j_{q,z}^{c} = -\lambda \frac{\partial T}{\partial z} \tag{2.51}$$

Heat transport due to species diffusion is included in Eq. (2.28) via the term $\sum_{i=1}^{N_g} c_{p,i}\vec{j}_i \cdot \nabla T$. Dufour effect indicates the heat transport due to concentration gradients, which is a reciprocal process of thermal diffusion [45],

$$\vec{j}_q^{D} = \sum_{i=1}^{N_g}\sum_{\substack{j=1 \\ j\neq i}}^{N_g} \frac{cRTX_iX_j}{\rho_i}\frac{D_i^T}{D_{ij}}\left(\frac{\vec{j}_i}{\rho_i} - \frac{\vec{j}_j}{\rho_j}\right) \quad . \tag{2.52}$$

Duffour effect is neglected in this study.

2.1.2.3 Momentum Transport

In the momentum equation (Eq. (2.13)), momentum flux vector appears as the divergence of the stress tensor T [40]

$$\vec{j}_V = \nabla \cdot \mathrm{T} \quad . \tag{2.53}$$

Dynamic viscosity μ, which appears in the stress tensor, is calculated from the viscosity of each individual species in the mixture according to empirical approximation [30]

$$\mu = \frac{1}{2}\left[\sum_i X_i\mu_i + \left(\sum_i \frac{X_i}{\mu_i}\right)^{-1}\right] \tag{2.54}$$

where μ_i is the viscosity of the species i. It is calculated from the standard kinetic theory expression

$$\mu_i = \frac{5}{16} \frac{\sqrt{\pi m_i k_B T}}{\pi \sigma^2 \Omega^{(2,2)*}} \tag{2.55}$$

where m_i is the mass of the molecule i, $\Omega^{(2,2)*}$ is the collision integral, σ is the collision diameter and k_B is the Boltzmann constant, as it was explained in [40, 46].

2.1.3 Thermochemistry of the Gas-phase

In the conservation equations, some thermodynamic parameters are used to relate the heat with chemical and physical changes. In this case, enthalpy h and entropy s of the mixture and heat capacity of each species $c_{p,i}$ are calculated as a function of temperature and pressure. Specific enthalpy and entropy of an ideal mixture is defined as [47]

$$h = \sum_i Y_i h_i$$

$$s = \sum_i Y_i s_i \tag{2.56}$$

The change of the enthalpy and entropy of the chemical species i is calculated through the total differentials as

$$dh_i = \left(\frac{\partial h_i}{\partial T}\right)_p dT + \left(\frac{\partial h_i}{\partial p}\right)_T dp$$

$$ds_i = \left(\frac{\partial s_i}{\partial T}\right)_p dT + \left(\frac{\partial s_i}{\partial p}\right)_T dp \tag{2.57}$$

Specific enthalpy of the individual chemical species is independent from the pressure for ideal gases. Therefore partial differentials of Eq. (2.57) can be written as

$$\left(\frac{\partial h_i}{\partial T}\right)_p = c_{p,i} \ , \left(\frac{\partial h_i}{\partial p}\right)_T = 0 \ , \left(\frac{\partial s_i}{\partial T}\right)_p = \frac{c_{p,i}}{T} \ , \left(\frac{\partial s_i}{\partial p}\right)_T = -\frac{1}{\rho_i T} \quad . \tag{2.58}$$

In general it is not possible to determine the absolute values of enthalpy. However, enthalpy is a function of state, which means that the changes in enthalpy, ΔH, have absolute values. In this case, enthalpies of certain elements at specific temperature and pressure are set to zero, in which they are in their most stable form. Enthalpies of other substances are determined relative to this zero. *The standard enthalpy of formation* (ΔH_f°) of a substance is defined then as the enthalpy change in a reaction when one mole of a substance is formed in the standard state (298 K and 1 bar) from the reference forms of the elements in their standard states.

$$h_i(T) = h_{i,f}^0 + \int_{T^0}^{T} c_{p,i}(T)dT$$

$$s_i(T,P) = s_{i,f}^0 + \int_{T^0}^{T} \frac{c_{p,i}(T)}{T} dT + \int_{p^0}^{p} \frac{R}{pM_i} dp \tag{2.59}$$

Thermodynamic properties of the species i is calculated by a polynomial fit to fourth order to the specific heat at constant pressure

$$c_{p,i}(T) = a_1 + a_2 T + a_3 T^2 + a_4 T^3 + a_5 T^4 \tag{2.60}$$

where the coefficients a_1, a_2 etc. are taken from the experimental data.

2.2 Catalytic Surface

Every surface and every catalytic material have different properties. A catalytic surface is commonly characterized by its total site density Γ. Total site density is defined as the maximum number of sites available for adsorption per unit area of the surface. Its unit is given as (mol/m^2). It depends on the atomic arrangement of the catalytic surface. For instance, Γ is 2.49×10^{-5}, 1.53×10^{-5} and 2.16×10^{-5} mol/m^2 for 111, 110 and 100 platinum surfaces, respectively [48]. In addition, heterogeneous catalytic process modeling requires defining the state of the catalytic surface at a given period of time. In this respect, the state of a catalytic surface can be described by its coverages and temperature. Calculating the coverages and the catalyst temperature is explained later.

2.2.1 Thermochemistry of the Surface

Thermochemistry of the surface species is defined with the adsorption processes. Therefore, adsorption enthalpy and adsorption entropy are considered. In this respect, thermodynamic properties of the surface species can be principally calculated from the correlations that are given for gas-phase species. The transition state theory together with the statistical thermodynamics can be used for calculating the thermodynamic coefficients of the surface species. It is referred to [49, 50] for more detailed considerations such as heat of adsorption and entropy calculations.

2.3 Chemical Reactions

A chemical reaction is a process where one chemical substance is transformed to another through the rearrangement or exchange of atoms. Chemical reactions might occur at different phases: gas-phase, solid-phase, liquid-phase or a mixture of two different phases (solid + gas phase) [49]. In this respect, a homogeneous reaction undergoes over a single-phase. A heterogeneous reaction undergoes between different phases. In addition, every chemical reaction occurs at a certain rate. Therefore, chemical reactions are studied and investigated under reaction rate theories, which are derived for a particular reaction phase/or phases [49]. These reaction rate theories use certain parameters such as temperature, pressure, concentration of species, catalyst or inhibitor. As explained in the previous sections, chemically reacting flow over a catalytic surface requires considering the homogeneous gas-phase reactions and heterogeneous surface reactions. In the following three sections, fundamentals of reaction rate theories, homogeneous gas-phase and heterogeneous surface reactions are explained briefly.

2.3.1 Reaction Rate

A simple chemical reaction can be written as

$$\sum_{i=1}^{N} v_i' S_i \rightarrow \sum_{i=1}^{N} v_i'' S_i \tag{2.61}$$

where S is a reactant or product, and v_i' and v_i'' are the stoichiometric coefficients. As mentioned above, each reaction occurs at a particular rate. The rate of a chemical reaction can be described as the rate of the consumption of reactants, or rate of the creation of products. It is dependent on the concentrations of the reactants and the temperature. The rate of the formation or consumption in a reaction can be written then as [51]

$$RR = k \prod_{i=1}^{N} (c_i)^{v_i'} \tag{2.62}$$

where k is the reaction rate constant and c_i is the concentration of species i. k is dependent on the temperature, and can be written as

$$k = AT^{\beta} exp(-\frac{E_a}{RT}) \tag{2.63}$$

where AT^{β} is the collision frequency, and E_a/RT is the Boltzmann factor.

The pre-exponential factor (A), temperature exponent (β) and activation energy (E_a) are independent of the concentrations and temperature. Considering Eq. (2.62) and Eq. (2.63) together gives the net reaction rate, and it can be written for a single reaction as

$$\frac{dc_i}{dt} = (v_i'' - v_i')\, k \prod_{i=1}^{N} c_i^{v_i'} \tag{2.64}$$

2.3.2 Global and Elementary Reactions

In chemical kinetics, reactions are generally investigated as global (overall) or elementary reactions. Global reactions usually have very complicated rate laws. Their kinetic data is derived from the experiments by measuring the temperature, pressure and species concentrations [52]. Therefore, global reactions are applicable only for the measured conditions. In addition, detailed investigations have shown that reactions are formed via elementary steps rather than a single step. For instance, water is formed via different elementary steps, where different intermediate radicals or molecules are formed [53]. Therefore, every reaction, simple or complex, heterogeneous surface reactions or homogeneous gas-phase reactions, can be investigated in elementary steps. In addition, the coefficients in the Arrhenius equation have physical meanings, which can be measured [54]. Therefore, elementary reactions give the possibility to investigate the interactions between the reacting species on a molecular level over a wide range of temperature and pressure conditions. In elementary reactions, reaction molecularity is defined as the number of reactants that involve for the production of products. There can be three different reaction molecularity, i.e., unimolecular (single reactant molecule), bimolecular (two reactant molecules) or termolecular (three reactant molecules). In this thesis, only the elementary reaction scheme is used.

2.3.3 Homogeneous Gas-phase Reactions

Homogeneous gas-phase reactions occur only in the gas-phase. They provide source or sink terms in the species continuity equations, and heat release in the energy equation in the gas-phase. Therefore, they must be included for modeling the reactive flows.

An irreversible simple elementary gas-phase reaction can be given with Eq. (2.61). Since chemical reactions are reversible, Eq. (2.61) should be expanded to include the backward reactions as well. The following equation can be written for a reversible reaction as

$$\sum_{i=1}^{N_g} v'_{ir}\, S_i \underset{k_{b,r}}{\overset{k_{f,r}}{\rightleftarrows}} \sum_{i=1}^{N_g} v''_{ir}\, S_i \qquad (r = 1, \dots, R)\,. \tag{2.65}$$

where r is the considered reaction and R is the total number of reactions. In this case, the rate-of-progress (mol/m^3·s), which is the difference between forward and backward reactions, can be written for the rth reaction as [55]

$$\dot{\omega}_r = k_{f,r} \prod_{i=1}^{N_g} (c_{S_i})^{v'_{ir}} - k_{b,r} \prod_{i=1}^{N_g} (c_{S_i})^{v''_{ir}}\,. \tag{2.66}$$

in which v'_{ir} and v''_{ir} are the stoichiometric coefficient of species i in reaction r. $k_{\mathrm{f},r}$ and $k_{\mathrm{b},r}$ are the forward and backward reaction rate constants respectively in reaction r, and they are calculated by using Eq. (2.63). Here, $\dot{\omega}_r$ can be positive or negative depending on whether the forward or backward reaction proceeds faster [55]. In chemical reactions, there can be an equilibrium point as well, when forward and backward reactions proceed at the same rate on a microscopic level. This phenomenon is called as 'chemical equilibrium'. Chemical equilibrium can be explained with equilibrium constant, which can be given in general form as

$$K_{c,r} = \frac{k_{f,r}}{k_{b,r}} \ . \tag{2.67}$$

Equilibrium constant can be investigated in detail by correlating the Helmholtz free energy (A), Gibbs free energy (G), and chemical potential (ζ) [47, 55]. In this respect, it can be given with respect to concentration and pressure as

$$K_{p,r} = exp\left(-\frac{\Delta_R \bar{G}^o}{RT}\right) \quad and \quad K_{c,r} = exp\left(-\frac{\Delta_R \bar{A}^o}{RT}\right)(c^0)^{\sum_{i=1}^{N_g} v_{ir}} \ . \tag{2.68}$$

2.3.4 Heterogeneous Surface Reactions

In homogeneous gas-phase reactions there exist a single phase, where the rate laws can be easily integrated. However, in heterogeneous surface reactions, there exist two different phases, where the catalyst is in solid form and the reactants and products are in gaseous form. Therefore, the interaction between the solid surface and the adjacent gas constitute a system of complex reactions.

Heterogeneously catalyzed gas-phase reactions can be described by the elementary reaction steps of the catalytic process, including adsorption, surface diffusion, chemical transformation of the adsorbed species, and desorption [52]. Molecular aspects of heterogeneous catalytic processes can be elucidated with different models with different complexities such as *Density Functional Theory* (*DFT*), *Kinetic Monte Carlo* (*kMC*) and *Molecular Dynamics* (*MD*). *DFT* simulations perform quantum chemical calculations to investigate the quantitative determination of the interactions between adsorbates, so-called lateral interactions, on transition surfaces [56]. *kMC* model enables taking into account the fluctuations, correlations and the spatial distribution of the reaction intermediates on the catalyst surface. Therefore, the interactions between the molecules, the diffusion of the intermediates onto the surface, and adsorption/desorption of the reactants/intermediates, including different site demands and the activation barriers for the elementary reaction steps can be simulated with *kMC* [57]. *MD* approach accounts for the trajectories of atoms representing the surface and gaseous colliders by integrating Newton's equation of motion. The model calculates the interatomic forces. In *MD*, initial conditions of the surface and the active molecules are specified, and the results of the simulations enlighten a microscopic view of the collision that leads to adsorption and surface-transformation dynamics [58]. In general, accounting for the effect of the lateral interactions of the adsorbates for calculating the reaction rates is a challenging task, and it is difficult to couple it with CFD for simulating practical reactor configurations. In this case, *mean-field approximation (MF),* which neglects the effect of the lateral interactions of the adsorbates and non-uniformity of the surface, is a frequently used micro-kinetic approach for calculating the surface reaction rates in analogy with gas-phase reactions, and coupling them with the CFD for simulating the behavior of the practical reactor configurations. In this thesis, only the *mean-field approximation* is used for calculating the surface reaction rates.

2.3.4.1 Mean-field Approximation

In the mean-field approximation, every gas-phase species that is adsorbed on the surface (adsorbate) and catalytically active solid adsorbent are defined as surface species. The coverage of a surface species is defined as

$$\theta_i = \frac{\text{Number of adsorption sites occupied by species } i}{\text{Total number of adsorption sites avaliable}} \quad .$$

In this model, it is assumed that adsorbates are randomly distributed over the catalyst surface. The temperature of the catalyst and the coverages therein depend on time and macroscopic position in the reactor, but they are averaged over microscopic local fluctuations [52]. Therefore, the surface is assumed to be uniform. A surface reaction can be given then in analogy to Eq. (2.65) as

$$\sum_{i=1}^{N_g+N_s+N_b} v'_{ir}\, S_i \;\rightarrow\; \sum_{i=1}^{N_g+N_s+N_b} v''_{ir}\, S_i \qquad (r = 1, \dots, R)\,. \tag{2.69}$$

where S_i can be now a gas-phase species, a surface species that is adsorbed on the top of the monoatomic layer of the catalytic particle or a bulk species in the inner solid catalyst [52]. N_g, N_s and N_b represent the total number of gas-phase, surface and bulk species, respectively. Here, it should be also taken into account that different species occupy different number of adsorption sites. For instance, larger molecules might occupy more than one sites on the surface. Therefore, each species is assigned a "site occupancy number", σ_i [49]. In a surface reaction, the total number of the surface sites should be constant for an adsorption process, a reaction between adsorbed reactants, a desorption process, or species diffusion into and from the bulk phase [59]:

$$\sum_{i=1}^{N_s} v_{ir}\sigma_i = 0 \tag{2.70}$$

where v_{ir} is the difference of the stoichiometric coefficients ($v_{ir} = v''_{ir} - v'_{ir}$). It is now possible to derive the molar production rate $\dot{s}_i$ of a gaseous species, an adsorbed surface species or a bulk species due to surface reactions, under given assumptions, in analogy to Eq. (2.66) as

$$\dot{s}_i = \sum_{r=1}^{R} v_{ir}\, k_{f,r} \prod_{j=1}^{N_g+N_s+N_b} c_j^{v'_{jr}} \tag{2.71}$$

where *r* is the considered reaction and R is the total number of surface reactions. c_j is here the concentration of the species *j*, which is given in mol/m^2 for the adsorbed species and mol/m^3 for gaseous and bulk species

[59]. The general Arrhenius equation Eq. (2.63) can also be used for calculating the reaction rate coefficient $k_{\mathrm{f},r}$. However, rate constants should be modified with the coverages of the surface species. Because the binding states of the adsorption of all species vary with the surface coverages. Therefore, pre-exponential factor and the activation energy are written in rate coefficient as functions of the surface coverage of any surface species, as follows [49, 52]:

$$k_{f,r} = A_r T^{\beta_r} exp\left(-\frac{E_{a_r}}{RT}\right) \prod_{i=1}^{N_s} \theta_i{}^{\mu_{ir}} exp\left[\frac{\varepsilon_{ir}\theta_i}{RT}\right] \tag{2.72}$$

where A_r and β_r are the pre-exponential factor and temperature exponent, and μ_{ir} and ε_{ir} are the coverage parameters for species i, in reaction *r*, respectively.

For a reversible reaction, the forward and backward reaction rate constants are related through the equilibrium constant as given in Eq. (2.67). The unit of $K_{c,r}$ is given here in terms of concentration, however, it is more convenient to determine the equilibrium constant from the thermodynamics properties in pressure units, $K_{p,r}$, as follows [49]:

$$K_{c,r}(T) = K_{p,r}(T)\left(\frac{p^o}{RT}\right)^{\sum_{i=1}^{Ng} v_{ir}} \prod_{i=1}^{N_s}\left(\frac{\Gamma}{\sigma_i}\right)^{v_{ir}} \tag{2.73}$$

where p^o is the standard pressure at 1 bar. The equilibrium constant $K_{p,r}$ is calculated as

$$K_{p,r} = exp\left(\frac{\Delta S_i^o}{R} - \frac{\Delta H_i^o}{RT}\right) \tag{2.74}$$

where Δ is referring the change that occurs in passing completely from reactants to products in the *r*th reaction [49]:

$$\frac{\Delta S_i^o}{R} = \sum_{i=1}^{N} v_{ir} \frac{S_i^o}{R} \tag{2.75}$$

$$\frac{\Delta H_i^o}{RT} = \sum_{i=1}^{N} v_{ir} \frac{H_i^o}{RT} \tag{2.76}$$

2.3.4.2 Calculation of the Surface Coverages

Surface coverage of the ith species is calculated from the relationship between its concentration and site occupancy number, and surface site density as

$$\theta_i = \frac{c_i \sigma_i}{\Gamma} \quad . \tag{2.77}$$

Temporal variation of the coverage of the ith species is given as

$$\frac{\partial \theta_i}{\partial t} = \frac{\dot{s}_i \sigma_i}{\Gamma} \quad . \tag{2.78}$$

The sum of coverages should fulfill the following condition

$$\sum_{i=1}^{N_s} \theta_i = 1 \tag{2.79}$$

2.3.4.3 Sticking Coefficient

The sticking coefficient S_i^o can be defined as the ratio of the particles that are probably adsorbed on the surface to the total number of the particles that impinge upon the surface at a particular period of time. This probability ($0 \leq S_i^o \leq 1$) considers the existence of the suitable adsorption sites and influence of the lateral interactions between other adsorbed species [60]. It might be highly temperature dependent, and the temperature dependence of the sticking coefficient is given in analogy to Arrhenius expression as [49]

$$S_i^0 = a_i T^{b_i} e^{-\tilde{c}_i/RT} \tag{2.80}$$

where a_i and b_i are unitless and c_i has units compatible with the gas constant R. The local adsorption probability can defined then as

$$S_i^{eff} = S_i^0 \prod_{j=1}^{N_s} \theta_j^{\nu'_{jr}+\mu_{jr}} \tag{2.81}$$

The reaction rate of species i, $\dot{s}_i$, can be calculated from the kinetic theory of gases as

$$\dot{s}_i = S_i^{\mathrm{eff}} \sqrt{\frac{\mathrm{R}T}{2\pi M_i}} c_i$$

2.4 Modeling Mass Transport in the Washcoat and Coupling it with Surface Reactions

As mentioned in the introduction section, inclusion of a porous layer, called washcoat, over the solid support of the catalyst is a common application in heterogeneous catalytic processes. The purpose of including a porous washcoat structure over the catalyst support is increasing the surface area of the catalyst. In Fig. 2.3, an example is shown for a single quadratic channel of a honeycomb catalyst.

Figure 2.3: Scanning electron micrograph of a $CeO_2/Al2O_3$ washcoat in a 400 cpsi cordierite monolith, the figure is taken from [61]

In practical applications, the thickness of the washcoat can be as thin as 10-20 µm or as thick as 150-200 µm. In this case, heterogeneous chemical processes in the porous catalyst can be taken into account by considering two different approaches: 1) instantaneous diffusion, 2) finite diffusion through the catalyst. Instantaneous diffusion neglects the influence of the mass transport limitations on the reactant conversion in the catalyst. However, transport of chemical species in the washcoat and their surface reactions therein can be crucial for the overall behavior of the catalytic process, and consequently for the performance of the catalytic reactors. For instance, at high temperatures, when the rate of diffusion velocity is slower than the intrinsic reaction rate, reactants' concentrations in the catalyst decrease along its depth, and high concentration gradients occur in the catalyst. At low temperatures, when the intrinsic rate of reaction is slower than the diffusion velocity, concentration gradients in the catalyst become smaller. In addition, convective flow can also play significant role on mass transport in certain applications. In this case, the transport models, which are coupled with the surface reactions, should be comprehensive enough to estimate these effects. For this purpose, physical properties of the washcoat such as its thickness and porosity, and diameters of the inner pores should be incorporated into the transport models.

In this section, various transport models, from simple to detailed, with surface reactions in the washcoat are mentioned.

2.4.1 Instantaneous Diffusion (*∞-approach*)

Instantaneous diffusion model assumes that the catalyst is virtually distributed at the gas/washcoat interface, so that there is infinitely fast mass transport within the washcoat. This model neglects the washcoat parameters, such as its thickness and porosity, and the diameters of the inner pores. Therefore, *∞-approach* does not account for internal mass transport limitations that are due to a porous layer. It means that mass fractions of gas-phase species on the surface are obtained by the balance of production or depletion rate with diffusive and convective processes [20].

2.4.2 Effectiveness Factor Approach (*η-approach*)

Effectiveness factor approach accounts for diffusion limitations in the washcoat. *η-approach* is based on the assumption that one target species determines overall reactivity [5]. An effectiveness factor for a first order reaction is calculated for the chosen species based on the dimensionless Thiele modulus (Φ) [62, 63], and all reaction rates are multiplied by this factor at the species governing equation at the gas-surface interface. Φ is calculated as

$$\Phi = L\sqrt{\frac{\dot{s}_i\,\gamma}{D_{i,eff}\,c_{i,0}}} \tag{2.83}$$

in which $c_{i,0}$ is the concentration of species i at the gas-washcoat interface. γ in Eq. (2.83) stands for the active catalytic surface area per washcoat volume as

$$\gamma = \frac{F_{cat/geo}\,A_{geo}}{A_{geo}\,L} = \frac{F_{cat/geo}}{L} \tag{2.84}$$

in which L is thickness of the washcoat and $F_{\mathrm{cat/geo}}$ is the ratio of the total catalytically active surface area to the geometric surface area of the stagnation disc. Effective diffusion coefficient ($D_{i,\mathrm{eff}}$) calculation is referred to Eq. (2.94). The term in the square root in Eq. (2.83) indicates the ratio of intrinsic reaction rate to diffusive mass transport in the washcoat. When Thiele modulus is large, internal mass transfer limits the overall reaction rate; when Φ is small the intrinsic surface reaction kinetics is usually rate limiting [3].

Consequently, the effectiveness factor (η) is defined as the ratio of the effective surface reaction rate inside the washcoat to the surface reaction rate without considering the diffusion limitation [3]:

$$\eta = \frac{\dot{s}_{i,eff}}{\dot{s}_i} = \frac{\tanh(\Phi)}{\Phi} \tag{2.85}$$

Implementation of the η on the boundary conditions is given in Eq. (2.109). The zero-dimensional *η-approach* offers a simple and computationally inexpensive solution. However, it might lose the validity in conditions where more than one species' reaction rate and diffusion coefficient determines the overall reactivity.

2.4.3 Reaction-diffusion Equations (*RD-approach*)

Reaction-diffusion equations *(RD-approach)* offer a more adequate model than the *η-approach* to account for mass transport in the washcoat. The model calculates spatial variations of concentrations and surface reaction rates inside the washcoat. It assumes that the species flux inside the pores is only due to diffusion [64]. Therefore, it neglects the convective fluid flow inside the porous layer, because of very low permeability assumption [64]. Eventually, each gas-phase species leads to one reaction-diffusion equation in the *RD-approach*, which is written in the transient form, as

$$\frac{\partial c_{i,w}}{\partial t} = -\nabla \cdot \vec{j}_i^{\,w} + \gamma \dot{s}_{i,w} \tag{2.86}$$

$$\vec{j}_i^{\,w} = -D_{i,eff} \nabla c_{i,w} \tag{2.87}$$

in which $c_{i,w}$ is the molar concentration, $\vec{j}_i^{\,w}$ is the molar diffusion flux and $\dot{s}_{i,w}$ is the surface reaction rate of the ith species in the washcoat, respectively [20]. $\vec{j}_i^{\,w}$ is given for the cylindrical coordinates as

$$j_{i,r}^{w} = -D_{i,eff} \frac{\partial c_{i,w}}{\partial r} \tag{2.88}$$

$$j_{i,\theta}^{w} = -D_{i,eff} \frac{1}{r} \frac{\partial c_{i,w}}{\partial \theta} \tag{2.89}$$

$$j_{i,z}^{w} = -D_{i,eff} \frac{\partial c_{i,w}}{\partial z} \tag{2.90}$$

Effective Diffusion Coefficients in the Washcoat

η-approach and *RD-approach* models incorporate the physical parameters of the washcoat, i.e., the washcoat thickness, pore diameter, tortuosity and porosity, via effective diffusion coefficients. Effective diffusion coefficients are calculated from the molecular and Knudsen diffusion coefficients. Pore diameter determines if the molecular diffusion or Knudsen diffusion is more effective in the washcoat. In this respect, pore diameter in the washcoat can be classified as micropore, mesopore or macropore (Table 2.1).

Table 2.1: Pore diameters in the washcoat

Pore type	Pore diameter
Micropore	< 2 nm
Mesopore	2-50 nm
Macropore	> 50 nm

If the mean free path of the gaseous species is smaller than the mean pore diameter, the transport in the washcoat occurs mainly due to intermolecular collisions. Therefore, diffusion process is determined by the Fick's law, and the effective diffusion coefficients are calculated from the averaged molecular diffusion coefficients [59]:

$$D_{i,eff} = \frac{\varepsilon}{\tau}\left(D_{i,M}\right) \tag{2.91}$$

where ε is the washcoat porosity and τ is the tortuosity of the pores [27]. ε describes here the ratio of the void volume to the total volume of the washcoat. τ describes the longer connecting path imposed by obstacles within the washcoat relative to that for motion in unconstrained free path [65]. Washcoat porosity and tortuosity can be obtained through the experiments.

If the pore diameter is sufficiently small and the pressure in the washcoat is low, the mean free path of the gaseous species becomes larger than the pore diameter. In this case, the molecules collide with the walls of the washcoat more often than they collide with each other. This regime of mass transport in the washcoat is called 'Knudsen diffusion'. The Knudsen diffusion coefficient of the ith species is calculated as

$$D_{i,Knud} = \frac{d_p}{3}\sqrt{\frac{8RT}{\pi M_i}} \tag{2.92}$$

in which d_p is the mean pore diameter. Effective diffusion coefficients based on the Knudsen diffusion can be calculated then as

$$D_{i,eff} = \frac{\varepsilon}{\tau}\frac{d_p}{3}\sqrt{\frac{8RT}{\pi M_i}} \quad . \tag{2.93}$$

If the ratio of the mean free path to the mean pore diameter is close to unity, both molecular and Knudsen diffusion have to be taken into account. In this case, effective diffusion coefficients are calculated as

$$\frac{1}{D_{i,eff}} = \frac{\tau}{\varepsilon}\left(\frac{1}{D_{i,M}} + \frac{1}{D_{i,Knud}}\right) \tag{2.94}$$

2.4.4 Dusty-gas Model (*DGM*)

The Dusty-gas model takes the mass transport equations a step further by including the convective transport effect. The term dusty-gas is used for the fact that porous medium consists of large 'dust' molecules fixed in space. In this case, the pore walls are considered as large dust molecules, which are treated as a component of the gaseous mixture [66]. The kinetic theory of gases is applied then to this dusty-gas mixture.

In *DGM*, species transport inside the washcoat accounts for ordinary and Knudsen diffusion as well as the pressure-driven convective flow (Darcy flow) [67, 68]. The species mass conservation inside the washcoat is given in a conservative form as

$$\varepsilon\frac{\partial(\rho_g Y_i)}{\partial t} = -\nabla\cdot\vec{j}_i^{DGM} + \gamma\dot{s}_i M_i \tag{2.95}$$

Total mass density inside the washcoat is given as

$$\varepsilon\frac{\partial(\rho_g)}{\partial t} = -\sum_{i=1}^{N_g}\nabla\cdot\vec{j}_i^{DGM} + \sum_{i=1}^{N_g}\gamma\dot{s}_i M_i \tag{2.96}$$

In *DGM*, the fluxes of each species are coupled with one another [69]. The species molar fluxes are evaluated here using *DGM* as it is given in [70]

$$\vec{j}_i^{DGM} = -\left[\sum_{n=1}^{N_g} D_{i,n}^{DGM}\nabla c_n + \left(\sum_{n=1}^{N_g}\frac{D_{i,n}^{DGM}c_n}{D_{n,Knud}}\right)\frac{B_g}{\mu_w}\nabla p_w\right] \tag{2.97}$$

where c_n is the concentration of the nth gas-phase species, and μ_{w} is the viscosity of the mixture in the washcoat. Species fluxes in *DGM* are given for cylindrical coordinates as

$$\mathrm{j}_{i,r}^{\mathrm{DGM}} = -\left[\sum_{n=1}^{N_\mathrm{g}} D_{i,n}^{\mathrm{DGM}}\frac{\partial c_n}{\partial r} + \left(\sum_{n=1}^{N_\mathrm{g}}\frac{D_{i,n}^{\mathrm{DGM}}c_n}{D_{n,\mathrm{Knud}}}\right)\frac{B_\mathrm{g}}{\mu_\mathrm{w}}\frac{\partial p_\mathrm{w}}{\partial r}\right] \tag{2.98}$$

$$j_{i,\theta}^{DGM} = -\left[\sum_{n=1}^{N_g} D_{i,n}^{DGM}\frac{1}{r}\frac{\partial c_n}{\partial\theta} + \left(\sum_{n=1}^{N_g}\frac{D_{i,n}^{DGM}c_n}{D_{n,Knud}}\right)\frac{B_g}{\mu_w}\frac{1}{r}\frac{\partial p_w}{\partial\theta}\right] \tag{2.99}$$

$$j_{i,z}^{DGM} = -\left[\sum_{n=1}^{N_g} D_{i,n}^{DGM}\frac{\partial c_n}{\partial z} + \left(\sum_{n=1}^{N_g}\frac{D_{i,n}^{DGM}c_n}{D_{n,Knud}}\right)\frac{B_g}{\mu_w}\frac{\partial p_w}{\partial z}\right] \tag{2.100}$$

In *DGM*, pressure (p_{w}) inside the washcoat is calculated from the ideal gas law. $D_{i,n}^{\mathrm{DGM}}$ in Eq. (2.97) is the matrix of diffusion coefficients. Diffusion coefficients ($D_{i,n}^{\mathrm{DGM}}$) can be calculated from the inverse matrix [67]:

$$D_{i,n}^{DGM} = H^{-1} \tag{2.101}$$

where the elements of the H matrix are determined as [67].

$$h_{i,n} = \left[\frac{1}{D_{i,Knud}} + \sum_{M \neq i} \frac{X_M}{D_{i,M}}\right] \delta_{in} + (\delta_{in} - 1)\frac{X_i}{D_{i,n}} \tag{2.102}$$

where $D_{i,\mathrm{Knud}}$ is the Knudsen diffusion coefficient of ith species as determined in Eq. (2.102). The permeability in Eq. (2.97) is calculated from the Kozeny-Carman relationship [70] as

$$B_g = \frac{\varepsilon^3 d_{pt}^2}{72\tau(1-\varepsilon)^2} \tag{2.103}$$

where d_{pt} is the particle diameter.

2.5 Modeling Heat transport and Coupling with Surface Reactions in the Washcoat

As mentioned in section 2.4, the thickness of the washcoat can vary between 10-200 µm. Since this is a relatively thin layer and the heat conductivity of the washcoat materials are high, washcoat is commonly treated as isothermal. However, there are also studies which treat washcoat as non-isothermal and solve the energy balance in it [64, 71]. In this case, an energy balance equation for the washcoat can be given as

$$\overline{\rho c_p} \frac{\partial T_s}{\partial t} = \lambda_{\mathrm{eff}} \nabla^2 T - \gamma \sum_{i=1}^{N_g+N_s} h_i \dot{s}_i M_i - \sum_{k=1}^{N_g} h_i \vec{j}_i^{\mathrm{w}} \tag{2.104}$$

where the left hand side represents the energy storage in the washcoat. The term $\overline{\rho c_p}$ is here the effective specific heat capacity of the combined washcoat and gas mixture in each cell of the washcoat [71]. The first term on the right hand side accounts for the conduction of the energy along the washcoat. Heat release due to surface reactions is modeled via the second term. The last term on the right hand side considers the heat transport due to species diffusion. Stutz et.al [64] have given the effective conductivity in the washcoat based on the variational approach which uses effective magnetic permeability of macroscopically homogeneous and isotropic multiphase materials [72]

$$\lambda_{eff} = \frac{1}{((1-\varepsilon)/3\lambda_{wc}) + (\varepsilon/(2\lambda_{wc} + \lambda_g))} - 2\lambda_{wc} \tag{2.105}$$

in which λ_{wc} is the thermal conductivity of the washcoat and λ_{g} is the thermal conductivity of the gas mixture in each cell of the washcoat [64].

2.6 Coupling of Chemically Reactive Flow with the Catalytic Disc / Washcoat

Coupling of chemically reactive flow with the catalytic disc / washcoat is accomplished through the boundary conditions, which are set at the gas-surface interface. When the species mass fractions at the gas-washcoat interface are calculated, a small gas-phase volume element, which is adjacent to the reactive solid surface, is considered together with a small washcoat volume element [73].

2.6.1 Species mass fraction at the gas-washcoat interface

It is possible to derive the mass fraction of a gas-phase species at the gas-washcoat interface by using Eq. (2.19). In this case, diffusive and convective processes as well as the production or depletion rate of species due to surface ($\dot{s}_i$) and gas-phase ($\dot{\omega}_i$) reactions are considered. As seen in Fig. 2.4, Ω^+ is the small volume element in the gas-phase. $\partial\Omega^+$ is the outer boundary of this volume element [73]. Similarly, Ω^- is the small volume element in the washcoat. $\partial\Omega^-$ is the outer boundary of this volume element. $\partial\Omega$ is the interface between the gas-phase and solid washcoat. $\vec{\psi}_f^+$ is the flux through the control surface of the gas-phase volume element. $\vec{\psi}_f^-$ is the flux through the control surface of the washcoat volume element. Consequently, the gas-phase species conservation equation at the gas-washcoat interface is given as

$$\int_{\partial\Omega} \rho \frac{\partial Y_i}{\partial t} dV = -\int_{\partial\Omega^+} \vec{\psi}_f^+ \vec{n}\, dA + \int_{\partial\Omega^-} \vec{\psi}_f^- \vec{n}\, dA + \int_{\Omega^+} \dot{\omega}_i M_i\, dV \qquad (2.106)$$

where the last term on the right hand side accounts for the production or depletion rate due to gas-phase reactions.

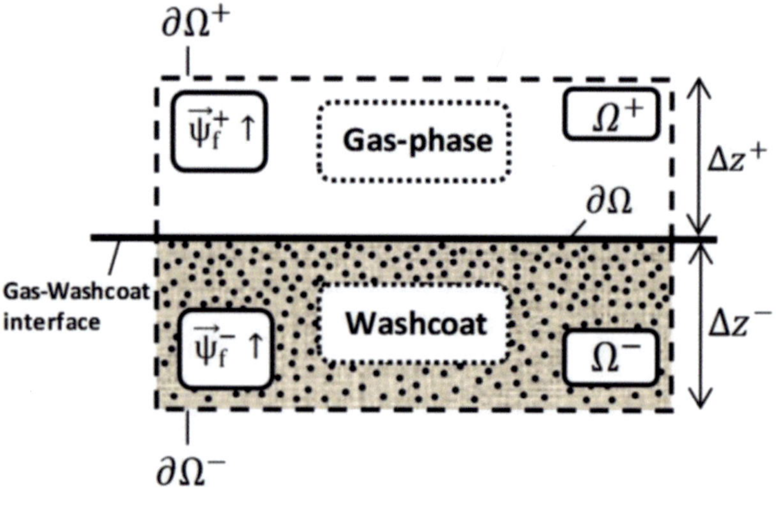

Figure 2.4: Gas-washcoat interface, small gas-phase and washcoat volume element adjacent to the interface

Eq. (2.106) can be rewritten by replacing the flux terms with the diffusion and convection processes and the surface reactions. In this respect, $\vec{\psi}_f^+$ is given as the sum of the convective and diffusive species fluxes from the gas-phase to the interface, i.e., $\vec{\psi}_f^+ = \vec{\mathrm{j}}_i + \rho Y_i \vec{\mathrm{u}}$. In case of an infinitely fast mass transport in the washcoat, $\vec{\psi}_f^-$ becomes the species flux due to adsorption and desorption at the gas-washcoat interface, i.e., $\vec{\psi}_f^- = \dot{s}_i M_i$. In this assumption, the diffusion in the washcoat is infinitely fast, but an important washcoat parameter $F_{\mathrm{cat/geo}}$, which is referring to the ratio of the catalytically active surface area to the geometric surface area of the stagnation disc, should also be accounted. Therefore, $\vec{\psi}_f^-$ flux term becomes $\vec{\psi}_f^- = F_{\mathrm{cat/geo}} \dot{s}_i M_i$. Consequently, Eq. (2.106) can be rewritten as

$$\int_{\partial\Omega} \rho \frac{\partial Y_i}{\partial t} dV = - \int_{\partial\Omega^+} (\vec{j}_i + \rho Y_i \vec{u}) \vec{n}\, dA + \int_{\partial\Omega} F_{cat/geo} \dot{s}_i M_i\, dA + \int_{\Omega^+} \dot{\omega}_i M_i\, dV \quad . \tag{2.107}$$

If chemical surface reactions occur, adsorption and desorption processes cause a net mass flow at the surface. This results in a flow velocity normal to the surface, which is called as Stefan-velocity ($\vec{u}$). It is calculated by summing the surface reaction rate of gas-phase species as

$$\vec{n}\, \vec{u} = \frac{1}{\rho} \sum_{i=1}^{N_g} \dot{s}_i M_i \tag{2.108}$$

In Eq. (2.107) internal mass transfer limitation in the washcoat is not accounted due to infinitely fast mass transport assumption. If internal mass transfer limitations in the washcoat are taken into account, the flux $\vec{\psi}_f^-$ is treated differently. In this case, three different approaches are considered for accounting for the internal mass transfer limitations, i.e., effectiveness factor approach, one dimensional (1D) reaction-diffusion equations and dusty-gas model.

If the *η-approach* is used, an effectiveness factor is multiplied with the surface reactions at the gas-washcoat interface. Therefore, Eq. (2.107) becomes,

$$\int_{\partial\Omega} \rho \frac{\partial Y_i}{\partial t} dV = -\int_{\partial\Omega^+} (\vec{J}_i + \rho Y_i \vec{u})\vec{n}\, dA + \int_{\partial\Omega} \eta F_{cat/geo} \dot{s}_i M_i\, dA + \int_{\Omega^+} \dot{\omega}_i M_i\, dV \quad . \tag{2.109}$$

If the *RD-approach* or *DGM* is used, the diffusion and reaction is calculated in the entire catalyst by resolving it in 1D. In this case, diffusion flux from the small washcoat element ($\vec{j}_i^{W}$) to the gas/washcoat interface is treated as an effective surface reaction rate, i.e., $\vec{\psi}_{\bar{f}} = \vec{j}_i^{W} M_i$. Therefore, Eq. (2.109) becomes,

$$\int_{\partial\Omega} \rho \frac{\partial Y_i}{\partial t} dV = -\int_{\partial\Omega^+} (\vec{J}_i + \rho Y_i \vec{u})\vec{n}\, dA + \int_{\partial\Omega} \vec{j}_i^{W} M_i\, dA + \int_{\Omega^+} \dot{\omega}_i M_i\, dV \quad . \tag{2.110}$$

2.6.2 Temperature at the gas/washcoat interface

Temperature of the catalyst is derived from various contributions of an energy balance between the solid support, washcoat and adjacent gas-phase. Figure 2.5 depicts the regarding volume elements.

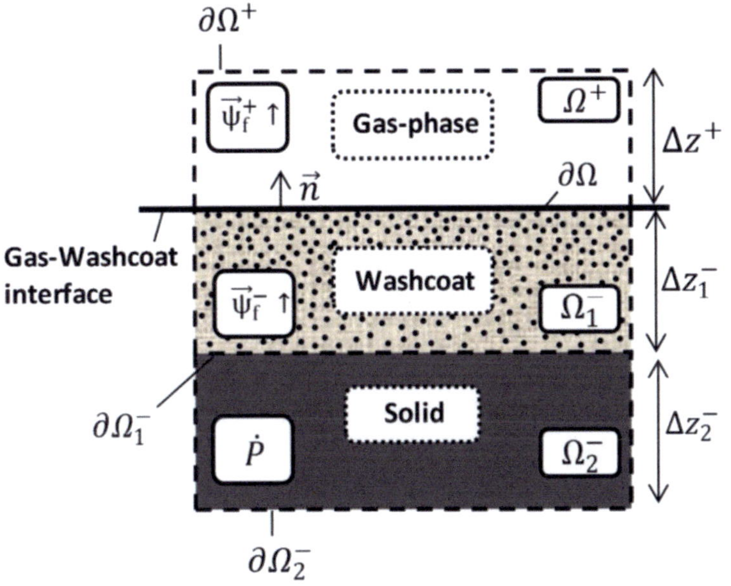

Figure 2.5: Thin solid and washcoated catalytic surface

For the energy balance, the conductive, convective and diffusive energy transport from the gas-phase to the surface, chemical heat release in the washcoat and in the adjacent gas-phase, the thermal radiation from the washcoat and from the solid support and resistive heating, conductive and convective energy losses of the solid support should be accounted.

If temperature gradient inside the washcoat is neglected, which means that the washcoat is isothermal, the solid support, washcoat and adjacent gas-phase will be in thermal equilibrium. In this case, the following energy equation can be written for the solid support, washcoat and adjacent gas-phase as

$$
\begin{aligned}
\int_{\Omega_1^-} \rho_{wc} c_{p,wc} \frac{\partial T}{\partial t} dV_{\Omega_2^-} &+ \int_{\Omega_2^-} \rho_s c_{p,s} \frac{\partial T}{\partial t} dV_{\Omega_1^-} + \int_{\Omega^+} \rho c_p \frac{\partial T}{\partial t} dV_{\Omega^+} \\
&= -\int_{\partial\Omega^+} \vec{j}_q^c \vec{n}\, dA \\
&+ \sum_{i=1}^{N_g} \int_{\partial\Omega^+} h_i(\vec{j}_i + \rho Y_i \vec{u})\vec{n} dA + \sum_{i=1}^{N_g} \int_{\Omega^+} \dot{\omega}_i M_i h_i dV \\
&- \sum_{i=1}^{N_g+N_s} \int_{\partial\Omega} \dot{s}_i M_i h_i dA - \int_{\partial\Omega} \vec{j}_{rad}\, \vec{n} dA - \int_{\partial\Omega_2^-} \vec{j}_{rad}\, \vec{n} dA + \dot{P} \\
&+ \int_{\Omega_2^-} \vec{j}_{loss}\, \vec{n} dA \quad .
\end{aligned}
\tag{2.111}
$$

where the first, second and third terms at the left-hand side represent the energy storage in the washcoat, solid support, and adjacent small gas-phase volume element, respectively. The first term on the right-hand side is the heat conduction from surface to gas according to the Fourier heat conductivity law. The second term accounts the convective and diffusive energy transport from the gas-phase to the surface. The third term is the heat release due to gas-phase reactions in the small gas-phase volume element. The fourth term describes the heat release due to surface reactions in the washcoat. The fifth and sixth terms are the heat radiation from the washcoat and solid support, respectively. And the last term on the right-hand side represents the conductive and convective energy losses of the solid support.

3 Numerical Modeling and Solution of the Stagnation-flow Reactor

The modeling approach of this thesis is based on the consideration of the SFR configuration (Fig. 1.2) in 1D. Evans and Greif [74] formulated a one-dimensional model of the rotating disk/stagnation-flow reactor. They considered two solid disks with a finite distance between them. Both disks had an infinite extent in the r-θ plane. In the rotating disk configuration, one of the disks was rotating, and the other parallel, porous disk was fixed. In the stagnation-point flow, both disks had a zero rotation rate. Gas at ambient temperature was injected through the porous disk normal to its surface. The rotating disk's surface was heated to a constant temperature. Coltrin et al. [75] extended the model to include the detailed chemical kinetics of species. Therefore, they included a species governing equation for each gas-phase species. These equations account for convective and diffusive transport of species, as well as production and consumption of species by elementary chemical reactions [75]. The CHEMKIN SPIN code [26], which was developed to solve 1D rotating-disk and SFR models, includes an equation for each surface species to consider the effect of surface composition on the system. The CHEMKIN SPIN code solves the models at steady-state. Deutschmann et al. [13] simulated the transient behavior at catalytic ignition with the 1D stagnation flow model. Raja et al. [25] formulated the compressible transient stagnation flowmodel to study the transient dynamics of catalytic ignition in stagnation flows.

In the following sections, initially steady-state 2D axisymmetric stagnation flow over a non-rotating surface is derived by considering 3D steady-state Navier-Stokes equations only in r-z coordinates, as it was given in [8].

Afterwards, the system is simplified further to 1D case based on the axi-symmetric stagnation flow equations. Further, the gas-phase equations are given in a form to use a transient iteration strategy to reach steady-state results. Since predicting the effect of internal mass transfer limitations in the SFR configuration is one of the main objectives in the present study, the model will be extended to include the diffusion limitations due to a porous catalytic layer.

3.1 Steady Axisymmetric Stagnation Flow Equations

Evans and Greif [74], Houtman et al. [7], Kee et al. [26, 76], Behrendt et al. [77], Deutschmann et al. [13] and Raja et al. [25] have formed the continuous development of the simplified formulations of the stagnation flows for semi-infinite and finite domains, steady and transient cases. Kee et al. [8] have documented all these cases comprehensively, which is also used as a main reference in this chapter. This subsection closely follows the explanations given in [8].

Axisymmetric stagnation flow equations are derived based on considering the steady-state 3D mass continuity and momentum equations. For the derivation, mass continuity and momentum equations equations are considered only in the r-z plane. Axisymmetric flow equations are valid for a certain regime, which is obtained from a relation between Rayleigh, Reynolds and Prandtl numbers. In axisymmetric flow configuration, variations of the variables with respect to the circumferential direction θ are not considered, therefore the derivatives with respect to θ drop out. A circumferential velocity component u_θ is only needed in *rotating surface* case. Therefore, circumferential momentum equation is also excluded here. In addition, bulk and dynamic viscosities are related through $\kappa = -2\mu/3$.

Further, two main conjectures are considered for the derivation. The first conjecture is based on considering the velocity field in terms of a stream function, which has a separable form

$$\psi(z,r) = r^2U(z) \tag{3.1}$$

where $U(z)$ is an unspecified function of z alone [8]. The advantage of the stream function is that it enables defining two different velocity variables in terms of a single variable. In addition, the axial momentum and mass continuity equations are combined into a single equation [7].

The second conjecture is based on presuming the changes in temperature, species composition and density in the z coordinate only [8]. Because, in the stagnation flow field, scalar quantities (temperature and species mass fractions) depend only on the distance from the surface, not on the radial position [19, 75, 77]. Under these assumptions, following flow equations are obtained as:

Mass continuity:

$$\frac{1}{r}\frac{\partial r\rho v_r}{\partial r} + \frac{\partial \rho v_z}{\partial z} = 0 \tag{3.2}$$

Axial momentum:

$$\rho v_r \frac{\partial v_z}{\partial r} + \rho v_z \frac{\partial v_z}{\partial z} = -\frac{\partial p}{\partial z} + \left[\frac{1}{r}\frac{\partial}{\partial r}\left(r\mu\left(\frac{\partial v_r}{\partial z} + \frac{\partial v_z}{\partial r}\right)\right) + \frac{\partial}{\partial z}\left(2\mu\frac{\partial v_z}{\partial z} - \frac{2}{3}\mu\nabla\cdot\vec{V}\right)\right] \tag{3.3}$$

Radial momentum:

$$\rho v_r \frac{\partial v_r}{\partial r} + \rho v_z \frac{\partial v_r}{\partial z} = -\frac{\partial p}{\partial r} + \left[\frac{\partial}{\partial r}\left(2\mu\frac{\partial v_r}{\partial r} - \frac{2}{3}\mu\nabla\cdot\vec{V}\right)\right.$$
$$\left. + \frac{\partial}{\partial z}\left(\mu\left(\frac{\partial v_r}{\partial z} + \frac{\partial v_z}{\partial r}\right)\right) + \frac{2\mu}{r}\left(-\frac{v_r}{r} + \frac{\partial v_r}{\partial r}\right)\right] \tag{3.4}$$

In the next step, partial derivative of the stream function with respect to *r* and *z* coordinates are considered as [8]

$$\frac{\partial \psi}{\partial r} = 2rU = r\rho v_z \tag{3.5}$$

$$-\frac{\partial \psi}{\partial z} = -r^2 \frac{dU}{dz} = r\rho v_r \tag{3.6}$$

Now some useful terms can be derived from Eq. (3.5) and Eq. (3.6), which can be used later to simplify the system further [8]:

$$2U = \rho v_z \quad , \quad -r\frac{dU}{dz} = \rho v_r \tag{3.7}$$

$$\frac{\partial v_z}{\partial z} = 2\frac{\partial}{\partial z}\left(\frac{U}{\rho}\right) \quad , \qquad \frac{\partial v_z}{\partial r} = 2U\frac{\partial}{\partial r}\left(\frac{1}{\rho}\right) = 0 \tag{3.8}$$

$$\frac{\partial v_r}{\partial z} = -r\frac{\partial}{\partial z}\left(\frac{1}{\rho}\frac{dU}{dz}\right), \qquad \frac{\partial v_r}{\partial r} = -\frac{dU}{dz}\frac{\partial}{\partial r}\left(\frac{r}{\rho}\right) = -\frac{1}{\rho}\frac{dU}{dz} \tag{3.9}$$

The divergence of the velocity in the fluid dilatation term can also be written as [8]

$$\nabla \cdot \vec{V} = \frac{\partial v_r}{\partial r} + \frac{\partial v_z}{\partial z} + \frac{v_r}{r} = 2\left(\frac{\partial}{\partial z}\left(\frac{U}{\rho}\right) - \frac{1}{\rho}\frac{dU}{dz}\right) \quad . \tag{3.10}$$

Inserting the relationships, which are derived in Eq. (3.7), Eq. (3.8), Eq. (3.9) and Eq. (3.10), into the momentum equations, eliminating the radial derivatives of the density or U (second conjecture: density is a function of *z* only) and isolating the pressure gradient terms on the left-hand sides of the momentum equations gives the following simplified differential equations [8]:

Axial momentum:

$$-\frac{\partial p}{\partial z} = 4U\,\frac{d}{dz}\left(\frac{U}{\rho}\right) + \frac{4}{3}\frac{d}{dz}\left[2\mu\frac{d}{dz}\left(\frac{U}{\rho}\right) + \frac{\mu}{\rho}\frac{dU}{dz}\right] - 2\mu\frac{d}{dz}\left(\frac{1}{\rho}\frac{dU}{dz}\right) \tag{3.11}$$

Radial momentum:

$$\frac{1}{r}\frac{\partial p}{\partial r} = 2U\,\frac{d}{dz}\left(\frac{1}{\rho}\frac{dU}{dz}\right) - \frac{1}{\rho}\left(\frac{dU}{dz}\right)^2 - \frac{d}{dz}\left[\mu\frac{d}{dz}\left(\frac{1}{\rho}\frac{dU}{dz}\right)\right] \tag{3.12}$$

3.2 Further Simplification to 1D Form

In Eq. (3.12) the radial pressure gradient is divided by *r* so that it can be written as a function of *z* only. In this case, the right hand sides of the axial and radial momentum equations will be functions of *z* only, and $\partial p/\partial z$ and $1/r(\partial p/\partial r)$ terms will also be functions of *z* only [8]. Differentiating the radial momentum equation once with respect to *z* and switching the differentiation order of the pressure with respect to *r* gives [8]

$$\frac{\partial}{\partial z}\left(\frac{1}{r}\frac{\partial p}{\partial r}\right) = \frac{1}{r}\frac{\partial}{\partial r}\left(\frac{\partial p}{\partial z}\right) = 0 \quad . \tag{3.13}$$

Eq. (3.13) involves that $1/r(\partial p/\partial r)$ should be constant [8]. This constant is denominated as the eigenvalue of the radial momentum equation Λ. Inserting the physical velocities back into the variable *U* and its derivative gives [8]

$$\rho v_z = 2U, \qquad \rho\frac{v_r}{r} = \rho V = -\frac{\partial U}{\partial z} \tag{3.14}$$

where $V = v_r/r$ is denominated as the scaled radial velocity. This new variable is also a function of *z* alone.

In the next step, thermal-energy and species-continuity equations are considered in 1D only with respect to z spatial coordinate (second conjecture: temperature and species-continuity are functions of z only) [8]. Perfect gas equation is included to close the equation system. Based on these considerations, 1D stagnation flow equations are obtained as

Mass continuity:

$$0 = -2\rho V + \frac{d(\rho v_z)}{dz} \tag{3.15}$$

Axial momentum:

$$\rho v_z \frac{dv_z}{dz} = -\frac{dp}{dz} + 2\mu \frac{dV}{dz} + \frac{4}{3}\frac{d}{dz}\left[\mu \frac{dv_z}{dz} - \mu V\right] + 2\mu \frac{dV}{dz} \tag{3.16}$$

Scaled radial momentum:

$$\rho v_z \frac{dV}{dz} + \rho V^2 = -\Lambda + \frac{d}{dz}\left(\mu \frac{dV}{dz}\right) \tag{3.17}$$

Thermal energy:

$$\rho v_z c_p \frac{dT}{dz} = -\sum_{i=1}^{N_g} j_{i,z}\, c_{p,i} \frac{dT}{dz} - \sum_{i=1}^{N_g} \dot{\omega}_i\, M_i h_i + \frac{d}{dz}\left(\lambda \frac{dT}{dz}\right) \tag{3.18}$$

Species continuity:

$$\rho v_z \frac{dY_i}{dz} = \dot{\omega}_i M_i - \frac{dj_{i,z}}{dz} \tag{3.19}$$

Perfect-gas equation:

$$p = \rho R T \sum_{i=1}^{N_g} \frac{Y_i}{W_i} \tag{3.20}$$

This simplified 1D SFR equations does not emerge due to neglecting certain physical effects, instead it emerges due to natural vanishing of some terms because of the mathematical reduction [8]. Therefore, it considers all certain physical and chemical effects, and it is convenient to investigate the gas-surface interactions at a detailed fundamental level. In this case, there are also other simplified models such as 1D plug flow and 2D boundary layer equations to predict the behavior of chemically reacting flows. These simplified models neglect some certain physical effects. For instance, plug flow reactor (PFR) model neglects radial gradients through the reactor [78]. In addition, convective transport is assumed to dominate over the diffusive transport in the axial direction [79]. These assumptions lead to a 1D model without considering any diffusive term. Boundary layer approximation ignores the diffusive transport terms along the flow direction and sets all the second derivatives involving in the flow direction to zero [79].

3.3 Finite-Gap Stagnation Flows on Porous Catalytic Surfaces

In this section, the mathematical model for the finite-gap stagnation flow over a porous catalytic surface (Fig. 3.1) is provided with the mass transfer in the porous catalytic layer and specific boundary conditions. In this study, the purpose of the finite-gap stagnation flow on porous catalytic surfaces is not finding transient results, instead using a transient iteration strategy to find steady-state results. The mathematical formulation of this strategy is explained in the following sections. The final mathematical form given in this section is used throughout the simulations.

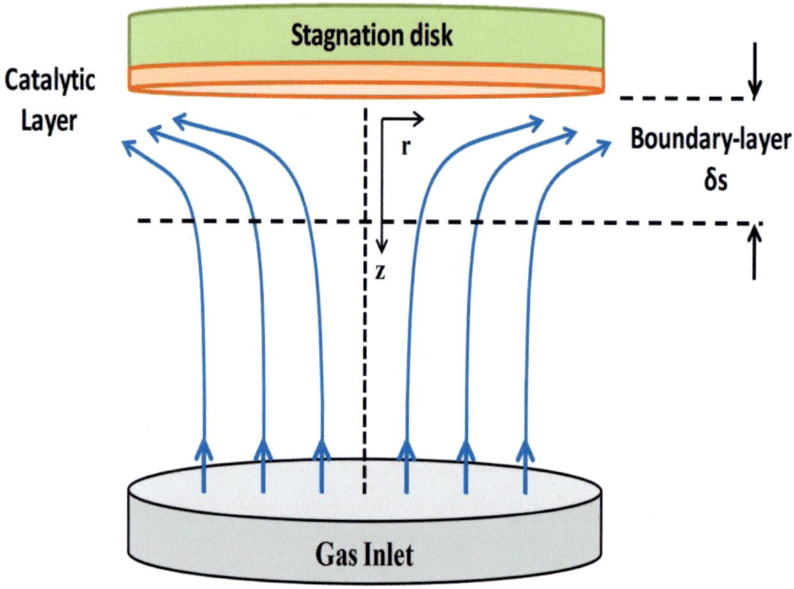

Figure 3.1: Schematic illustration of the stagnation flow configuration, the figure is taken from [20]

3.3.1 Gas Phase Equations

The gas-phase equations for the SFR, which are considered in this study, are based on Eq. (3.15)-Eq. (3.20). However, there are some alterations. Initially, compressible transient form of the stagnation flow equations is considered. In the transient formulation, the dependent variables are given with respect to time and axial coordinate as; axial velocity $v_z = v_z(t, z)$, scaled radial velocity $V = V(t, z)$, temperature $T = T(t, z)$, and species mass fraction $Y_i = Y_i(t, z)$ [80]. In addition, a physical characteristic length scale is introduced between incoming flow and stagnation surface due to finite-gap consideration, and pressure-curvature term Λ is solved as the eigenvalue of the system whose magnitude is adjusted to satisfy the remaining boundary conditions. In this case, eigenvalue of the momentum equations is given as $1/r(\partial p/\partial r) = \Lambda(t)$. Based on these considerations, the compressible stagnation flow equations are obtained as [80]

Mass continuity:

$$\frac{\partial \rho}{\partial t} = -2\rho V + \frac{\partial(\rho v_z)}{\partial z} \tag{3.21}$$

Axial momentum:

$$\rho \frac{\partial v_z}{\partial t} = -\frac{\partial p}{\partial z} - \rho v_z \frac{\partial v_z}{\partial z} + 2\mu \frac{\partial V}{\partial z} + \frac{4}{3}\frac{\partial}{\partial z}\left[\mu \frac{\partial v_z}{\partial z} - \mu V\right] + 2\mu \frac{\partial V}{\partial z} \tag{3.22}$$

Scaled radial momentum:

$$\rho \frac{\partial V}{\partial t} = \rho v_z \frac{\partial V}{\partial z} - \rho V^2 - \Lambda(t) + \frac{\partial}{\partial z}\left(\mu \frac{\partial V}{\partial z}\right) \tag{3.23}$$

Thermal energy:

$$\rho c_p \frac{\partial T}{\partial t} = \frac{\partial P}{\partial t} - \left[\rho c_p v_z + \sum_{i=1}^{N_g} \rho Y_k V_k c_{p,i}\right] \frac{\partial T}{\partial z} - \sum_{i=1}^{N_g} \dot{\omega}_i M_i h_i + \frac{\partial}{\partial z}\left(\lambda \frac{\partial T}{\partial z}\right) \tag{3.24}$$

Species continuity:

$$\rho \frac{\partial Y_i}{\partial t} = -\rho v_z \frac{\partial Y_i}{\partial z} + \dot{\omega}_i M_i - \frac{\partial \rho Y_k V_k}{\partial z} \tag{3.25}$$

Perfect-gas equation:

$$p = \rho R T \sum_{i=1}^{N_g} \frac{Y_i}{W_i} \tag{3.26}$$

As mentioned in section 3.3, the purpose of this study is not finding transient results. Therefore, further simplifications can be considered. The pressure variations are assumed to be small compared to the mean thermodynamic pressure [80]. Therefore, pressure in the system is assumed to be constant. Thermal energy equation is still considered in its transient form (Eq. (3.30)), only by excluding the transient pressure term. Species continuity equation is also given in its transient form (Eq. (3.31)). Continuity equation is still treated as an algebraic equation, but time derivative of mass density is included in the equation in terms of time derivative of species mass fractions and temperature (Eq. (3.27)). Axial momentum equation is decoupled from the equations, because it is not needed to determine the axial velocity v_z. Pressure-curvature term Λ is solved as the eigenvalue of the equation system again, but in its algebraic form. Under these considerations, the final form of the gas-phase equations emerge as [73]

Mixture continuity:

$$0 = \frac{p}{R}\frac{\bar{M}^2}{T^2}\left[T\sum_i \frac{\partial Y_i}{\partial t}\frac{1}{M_i} + \frac{\partial T}{\partial t}\frac{1}{\bar{M}}\right] - 2\rho V - \frac{\partial(\rho v_z)}{\partial z} \tag{3.27}$$

Radial momentum:

$$0 = -\frac{\rho v_z}{\rho}\frac{\partial V}{\partial z} - V^2 - \frac{\Lambda}{\rho} + \frac{1}{\rho}\frac{\partial}{\partial z}\left(\mu\frac{\partial V}{\partial z}\right) \tag{3.28}$$

Eigenvalue of the radial momentum:

$$0 = \frac{\partial \Lambda}{\partial z} \tag{3.29}$$

Thermal energy:

$$\frac{\partial T}{\partial t} = -\left[\frac{\rho v_z}{\rho} + \frac{1}{\rho c_p}\sum_{i=1}^{N_g} c_{p,i} j_i\right]\frac{\partial T}{\partial z} - \frac{1}{\rho c_p}\sum_{i=1}^{N_g} \dot{\omega}_i M_i h_i + \frac{1}{\rho c_p}\frac{\partial}{\partial z}\left(\lambda\frac{\partial T}{\partial z}\right) \tag{3.30}$$

Species continuity:

$$\frac{\partial Y_i}{\partial t} = -\frac{\rho v_z}{\rho}\frac{\partial Y_i}{\partial z} + \frac{1}{\rho}\dot{\omega}_i M_i - \frac{1}{\rho}\frac{\partial j_i}{\partial z} \tag{3.31}$$

Ideal gas law:

$$\rho = \frac{p\bar{M}}{RT} \tag{3.32}$$

In the governing equations, dependent variables of the system are the axial mass flux ρv_z, the scaled radial velocity V, the eigenvalue of the momentum equation Λ, the temperature T and the species mass fraction Y_i. Independent variables are the axial distance from the surface z and the time t. The axial mass flux ρv_z is considered as the dependent variable in the continuity equation (Eq. (3.27)), not only the axial velocity v_z, because axial momentum equation is already decoupled. The radial momentum equation is coupled to the continuity equation through the convection term.

3.3.2 Reaction and Diffusion in the Porous Catalyst of the SFR

Reaction and diffusion in the porous catalyst of the SFR is modeled as it was explained in section 2.4. Reaction-diffusion equations and dusty-gas model are considered only in 1D form with respect to the axial z coordinate only, which were explained in section 2.4.3 and 2.4.4, respectively.

3.3.3 Boundary Conditions

Boundary conditions are needed to close the equation system. In this case, the second-order flow equations require information on scaled radial velocity V, temperature T and species mass fractions Y_i for both inlet flow and stagnation surface. First-order continuity equation requires information from v_z on one boundary. There is no explicit boundary condition for Λ, but it must be provided in such a way that all other boundary conditions are satisfied [25]. The boundary conditions at the washcoat support side should also be included.

Inlet Boundary

Finite gap stagnation flow solution becomes relevant by introducing the inlet boundary conditions together with the physical characteristic length between the surface and gas-phase [8]. In this manner, boundary-layer thickness is small relative to the lateral extent of the reactor, and convection plays an important role in vertical transport of the momentum and mass. If the boundary-layer spans the whole chamber, convection plays a little role in the vertical transport of momentum and mass, and stagnation flow solution breaks down [81].

Dirichlet boundary conditions are considered for the temperature, mass fraction of each gas-phase species and scaled radial velocity at the inlet flow. The scaled radial velocity should fulfill the no-slip boundary condition.

$$T = T^0 \qquad Y_i = Y_{i,0} \tag{3.33}$$

$$V = 0 \tag{3.34}$$

The continuity equation at the inlet boundary is considered as a constraint equation and it is solved itself [25]. Therefore, an explicit boundary condition is not needed for the axial mass flux at the inlet boundary [8]. The following equation is considered for Λ at the inlet boundary,

$$F_{\Lambda,inlet} = \frac{(\rho v_x)_{g,N}}{(\rho)_{g,N}} - v_{x,0} \tag{3.35}$$

where (g, N) represents the grid point at the inlet side.

Gas-Surface/Washcoat Interface

In order to couple the outer surface and the surrounding flow, interaction between them must be considered as it is explained in section 2.6.1 and section 2.6.2. Therefore, energy balance and species conservation equations are established at the interface. In addition, the following integral relationship can be used for the small control volume element [73].

$$\int_{\Omega^+} dV = \Delta z^+ \int_{\partial\Omega^g} dA \tag{3.36}$$

Species governing equation at the interface can be written depending on the considered surface models by using Eq. (2.106), respectively:

∞-approach:

$$\rho \frac{\partial Y_i}{\partial t} \Delta z^+ = -j_i - \rho u Y_i + F_{cat/geo} \dot{s}_i \, M_i \tag{3.37}$$

η-approach:

$$\rho \frac{\partial Y_i}{\partial t} \Delta z^+ = -j_i - \rho u Y_i + \eta F_{cat/geo} \dot{s}_i \, M_i \tag{3.38}$$

RD-approach and DGM:

$$\rho \frac{\partial Y_i}{\partial t} \Delta z^+ = -j_i - \rho u Y_i - j_i^w \, M_i \tag{3.39}$$

In these equations Δz^+ is defined as the halfway between the gas-washcoat interface ($z_{g,1}$ in Fig.3.2) and the adjacent grid point in the gas-phase ($z_{g,2}$ in Fig.3.2). In addition, the effect of gas-phase reactions in the adjacent gas is excluded, because gas-phase reactions are not likely to occur at the temperature range considered in this thesis study. However, they can be included at the interface in case of high temperatures as it is given in Eq. (2.110).

Two different boundary conditions can be considered for the catalyst temperature. It can be either set to a constant surface temperature or calculated from an energy balance, i.e., from Eq. (2.111). In addition, the following integral relationships can be used for the washcoat and support,

$$\int_{\Omega_1^-} dV = \Delta z_1^- \int_{\partial\Omega} dA \quad and \quad \int_{\Omega_2^-} dV = \Delta z_2^- \int_{\partial\Omega_1^-} dA \tag{3.40}$$

The energy balance at the gas/washcoat interface can be given as

$$\begin{aligned}\left(\rho_{wc}c_{wc}\Delta z_1^- + \rho_s c_s \Delta z_2^- + \rho c_p \Delta z^+\right)\frac{\partial T}{\partial t} &= \lambda\frac{\partial T}{\partial z} \\ &- \sum_{i=1}^{N_g}(j_i + \rho Y_i u)h_i - \sigma\epsilon_{wc}\left(T^4 - {T_{rad}}^4\right) \\ &- \sigma\epsilon_s\left(T^4 - {T_{rad}}^4\right) - \sum_{i=1}^{N_g}\dot{s}_i M_i h_i + \dot{P} - \frac{\lambda_s}{\Delta z_2^-}(T - T_b)\end{aligned} \tag{3.41}$$

in which the first term on the right hand side of the equation accounts for heat conduction from the surface to the gas according to the Fourier heat conductivity law. λ is here the thermal conductivity of the gas, which is adjacent to the surface. The second term describes convective and diffusive energy transport from the gas-phase to the surface, where h_i is the enthalpy of species i. The third and fourth terms are heat radiation from the surface due to the Stefan-Boltzmann law, where σ is the Stefan-Boltzmann constant, $\varepsilon_{\mathrm{wc}}$ is the emissivity of the washcoat and ε_{s} is the emissivity of the support. Here T_{rad} is the reference temperature to which the surface radiates. The fourth term encompasses heat release due to chemical reactions. The fifth term contains the energy source corresponding to the resistive heating of the surface. The last term evaluates the conduction losses of the support. Here the temperature gradient inside the washcoat layer is neglected.

The continuity at the gas-surface/washcoat boundary is evaluated from,

$$F_{\rho v_z,surface} = (\rho v_z)_{g,1} \tag{3.42}$$

where the indices $g, 1$ represents the gas-washcoat interface location. The scaled radial velocity at the surface is specified as zero to fulfill the no-slip boundary condition due to the finite-gap case

$$V = 0 \quad . \tag{3.43}$$

The following equation is considered for Λ at the surface boundary [8]

$$F_{\Lambda,surface} = \Lambda_{g,2} - \Lambda_{g,1} \tag{3.44}$$

where *(g, 1)* and *(g, 2)* indices of Λ indicate the gas/washcoat interface and the adjacent grid point in the gas-phase, respectively.

Washcoat/Support Interface

For this boundary condition it is assumed that the washcoat is thick enough such that concentration gradients vanish at the washcoat/support boundary [82],

$$\left.\frac{\partial C_i}{\partial y}\right|_{y=\Delta z_1^-} = 0 \tag{3.45}$$

in which Δz_1^- is the thickness of the washcoat as it is also depicted in Fig. 2.5.

3.4 Numerical Solution of the Model Equations

For numerical solution, the partial-differential equations (PDE) regarding the gas-phase, washcoat, and boundary conditions are transformed to a system of ordinary differential and algebraic equations (DAE). This is accomplished by spatial discretization of the PDE system by using finite difference approximations on a non-equidistant grid.

3.4.1 Discretization of the Model Equations

The discretization scheme applied in the current study coherences in general with the discretization scheme of [8, 25]. However, the variables are always considered at the actual nodes. The discretization scheme is shown in Figure 3.2.

In Fig.3.2, $z_{g,1}$, $z_{g,2}$ and $z_{wc,1}$ represent the grid point at the gas-washcoat interface, the grid point in the gas-phase that is adjacent to the interface, and the first grid point in the washcoat, along the z direction, respectively. Similarly, $\Delta z_{g,1}$ and $\Delta z_{wc,1}$ represent the distances between two neighboring grid points along the z direction in the gas-phase and in the washcoat, respectively.

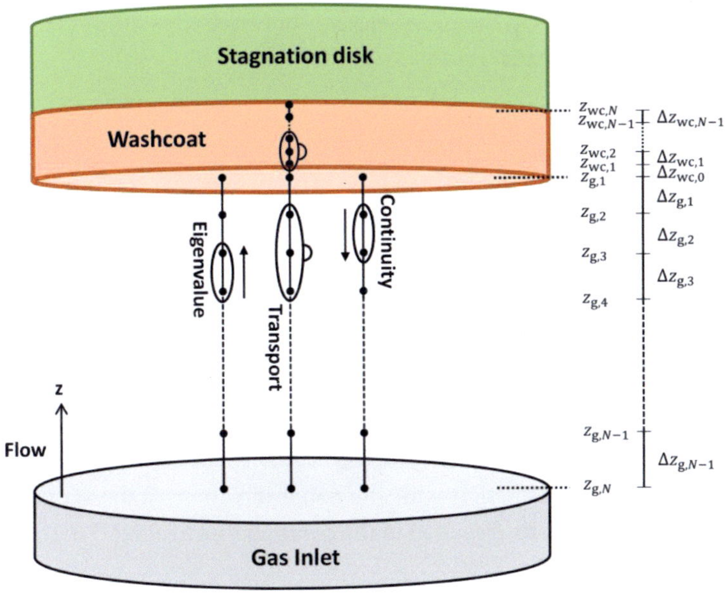

Figure 3.2: Spatial discretization scheme

The following central finite difference discretization scheme is used for the second order diffusive terms,

$$\frac{\partial}{\partial z}\left(f\frac{\partial g}{\partial z}\right)_l = \frac{1}{z_{l+1}-z_{l-1}}\left((f_{l+1}+f_l)\left(\frac{g_{l+1}-g_l}{z_{l+1}-z_l}\right)-(f_l+f_{l-1})\left(\frac{g_l-g_{l-1}}{z_l-z_{l-1}}\right)\right) \tag{3.46}$$

which correlates the variables at three neighboring grid points. For instance, the diffusive term in the energy equation is calculated at the actual grid point as,

$$\frac{\partial}{\partial z}\left(\lambda\frac{\partial T}{\partial z}\right)_l = \frac{1}{z_{l+1}-z_{l-1}}\left((\lambda_{l+1}+\lambda_l)\left(\frac{T_{l+1}-T_l}{z_{l+1}-z_l}\right)-(\lambda_l+\lambda_{l-1})\left(\frac{T_l-T_{l-1}}{z_l-z_{l-1}}\right)\right) \quad . \tag{3.47}$$

The convective term in the radial-momentum equation follows an upwind difference approximation as follows,

$$\frac{\rho v_z}{\rho}\frac{\partial V}{\partial z} = \frac{(\rho v_z)_l}{(\rho)_l}\frac{V_{l+1} - V_l}{z_{l+1} - z_l} \quad . \tag{3.48}$$

Convective term of the continuity equation is discretized in a way that it propagates the information from the lower boundary towards the inlet-boundary [8, 25],

$$\frac{\partial(\rho v_z)}{\partial z} = \frac{(\rho v_z)_l - (\rho v_z)_{l-1}}{z_l - z_{l-1}} \quad . \tag{3.49}$$

Discretization of the radial pressure gradient is applied in a way that it propagates the information with the same direction of the momentum transport and opposite direction of the continuity equation,

$$\frac{\partial \Lambda}{\partial z} = \frac{\Lambda_{l+1} - \Lambda_l}{z_{l+1} - z_l} \quad . \tag{3.50}$$

Grid Adaptation

Spatial discretization of the model equations is now straightforward. The equations are discretized initially on an equidistant mesh (coarse mesh). However, when the equations are solved, there can be sometimes high gradients between two adjacent grid points. For instance, there can be high temperature and species concentration gradients near the stagnation surface. Therefore, equidistant mesh approximation can give inadequate results. In this case, fine mesh usually gives more precise results. In this thesis, two different approximations are used for fine mesh generation. The first approximation is based on using a simple aspect ratio φ relative to one specific location in the reactor (gas-washcoat interface). This method creates finer mesh near the gas-washcoat interface. φ is here a pre-defined value. The mathematical formulation of this approximation follows:

$$m_1 = (\varphi_1)^0 + \sum_{j=2}^{g,N-1} \varphi_1^{j-1} \tag{3.51}$$

$$z_{g,1} = 0 \qquad 1st\ spatial\ grid\ point\ (g,1) \tag{3.52}$$

$$D = \frac{x_{len}}{m} \tag{3.53}$$

$$n_1 = D \tag{3.54}$$

$$z_{g,2} = z_{g,1} + n_1 \qquad 2nd\ spatial\ grid\ point\ (g,2) \tag{3.55}$$

$$\left.\begin{aligned} n_{l-1} &= n_{l-2} * \varphi_1 \\ z_{g,l} &= z_{g,l-1} + n_{l-1} \end{aligned}\right\} \quad the\ spatial\ grid\ points\ in\ between, \quad l = (g,3), \dots, (g,N) \tag{3.56}$$

where φ_1 is the aspect ratio for the gas-phase. $z_{g,1}$ is here the grid point at the gas-washcoat interface. Therefore, the value of $z_{g,1}$ should be zero. Similarly $z_{g,2}$ is the distance of the second grid point (in the gas-phase) from the surface and $z_{g,l}$ is the distance of the lth grid point (in the gas-phase) from the surface. x_{len} is here the total length of the finite-gap. Grid points in the washcoat can be generated by introducing another aspect ratio (φ_2) and using the Eq. (3.51)-Eq. (3.56). Grid generation based on aspect ratio is illustrated in Fig.3.3.

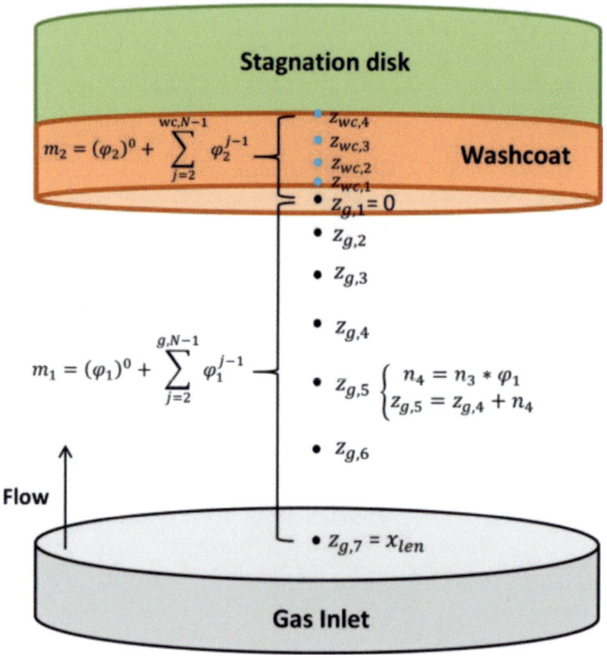

Figure 3.3: Fine mesh generation using aspect ratio

The second approximation is based on using an adaptive gridding, in which the necessary new grid points are automatically inserted into the coarse (equidistant) mesh points. In this respect, the adaptive gridding method, which was used in [26], is implemented here. The adaptive gridding is performed based on the following equations:

$$|\psi_{n,j} - \psi_{n,j-1}| \leq GRAD \times (max\ \psi_n - min\ \psi_n) \tag{3.57}$$

$$\left|\left(\frac{d\psi_n}{dz}\right)_j - \left(\frac{d\psi_n}{dz}\right)_{j-1}\right| \leq CURV \times \left(max\frac{d\psi_n}{dz} - min\frac{d\psi_n}{dz}\right) \quad . \tag{3.58}$$

In Eq. (3.57), the gradients are resolved by bounding the variation in the solution between mesh points. In Eq. (3.58), the curvature in the solution is resolved by bounding the variation in the solution's derivatives between mesh points. Therefore, these two expressions are calculated at each of the mesh points. When an inequality is not satisfied in a subinterval, a new mesh point is added automatically at the midpoint of the subinterval [26].

3.4.2 Differential Algebraic Equation System and Index Number

Discretization of the SFR model equations results in a system of differential algebraic equation (DAE). The general form of a DAE is given as

$$F(t, x, \dot{x}) = 0 \tag{3.59}$$

where F and x are vector values [83]. The explicit form of Eq. (3.59) can be given as

$$\dot{x} = f(t, x) \tag{3.60}$$

where the derivative of the dependent variable $\dot{x}$ is expressed explicitly with respect to the independent variable t and the dependent variable x. However, in a DAE system, derivatives of some of the dependent variables may not be expressed explicitly, namely they may not appear in the equations [84]. This case can be discussed with a restricted class of DAEs called semi-explicit nonlinear DAE, which is represented as

$$\begin{aligned} a' &= f(t, a, b) \\ 0 &= g(t, a, b) \end{aligned} \tag{3.61}$$

where the dependent variable vector is defined with the following transpose array $y = (a, b)^T$. It can be seen in Eq. (3.61) that some dependent variables (a: differentiable variables) have time derivatives, whereas others (*b*: algebraic variables) do not. In this case, the DAE system can be converted to ordinary differential equation (ODE) system by differentiating it with respect to the independent variable t. The *index* of the DAE system is expressed as the number of differentiation of the DAE to get a system of ODE [84]. If the following requirement is fulfilled, index of the DAE becomes 1:

$$\left|\frac{\partial g}{\partial b}\right| \neq 0 \tag{3.62}$$

which simply means that the determinant of the partial derivative of the algebraic equation *g* with respect to algebraic variable *b* should be nonsingular. This means that the algebraic constraints can in principle be solved for *g* in terms of *a* and *t*. There are many powerful numerical tools such as Sundials [85], Limex [86], DASSL [87], Matlab DAE Solver [88] and Twopnt [89] that can handle DAE index problem of 1. If the condition in Eq. (3.62) is not fulfilled, the determinant becomes singular. In this case, index of DAE becomes 2 or even higher. This problem occurs due to the reason that some of the algebraic variables define algebraic constraints between the differentiable variables only, rather than a relationship between the differentiable variables and the algebraic variables. In this respect, the index numbers of the discretized form of the SFR model should be analyzed whether this is the case. In the SFR model, three different discretized systems of equations emerge:

1) *system of equations emerging with ∞-approach and η-approach:*

   In *∞-approach* and *η-approach* washcoat is not spatially resolved. Therefore, surface reaction rates at the gas-washcoat interface are accounted as an implicit boundary condition on the system (Eq. (3.37) and Eq. (2.28)).

2) *system of equations emerging with the indirect coupling of RD-approach and DGM surface models with the surrounding flow:*

   In *RD-approach* and *DGM* surface models washcoat is resolved in 1D. However, in case of indirect coupling these surface models are not directly coupled to the flow equations, which means that in each new time step regarding the flow equations, *RD-approach* and *DGM* surface models are called on the gas-washcoat interface separately, and the concentrations at the gas-washcoat interface are passed to these surface models. These surface models are iterated separately until they reach steady state. After the surface models reach to their steady state conditions, effective surface reaction rates are transferred as an implicit boundary condition to the gas-washcoat interface.

3) *system of equations emerging with the direct coupling of RD-approach with the surrounding flow:*

   In case of direct coupling, the equations regarding the porous washcoat layer is solved simultaneously with the surrounding flow equations.

After classifying the different coupling strategies of the surface models, the dependencies of the model equations on different variables can be summarized in a similar scheme to [90] for these three different coupling strategies (the dependencies of the considered equations on the algebraic variables is shown with boldface):

1) *The dependencies of the model equations on different variables regarding the system of equations with ∞-approach and η-approach surface models:*

   surface equations

$$\frac{\partial \theta_i}{\partial t} = (\theta_i)_{i=1}^{N_s}\left(T^{g,1}, \left\{Y_j^{g,1}\right\}_{j=1}^{N_g}, \{\theta_k\}_{k=1}^{N_s}\right) \tag{3.63}$$

first spatial grid point (g, 1) (gas-washcoat interface)

$$\frac{\partial Y_i^{g,1}}{\partial t} = \left(T^{g,1}, T^{g,2}, Y_i^{g,1}, Y_i^{g,2}, \left\{Y_j^{g,1}\right\}_{j=1}^{N_g}, \{\theta_k\}_{k=1}^{N_s}\right) \tag{3.64}$$

$$\frac{\partial T^{g,1}}{\partial t} = \left(T^{g,1}, T^{g,2}, \left\{Y_j^{g,1}, Y_j^{g,2}\right\}_{j=1}^{N_g}, \{\theta_k\}_{k=1}^{N_s}\right) \tag{3.65}$$

$$0 = \left((\boldsymbol{\rho v_z})^{\boldsymbol{g,1}}, T^{g,1}, \left\{Y_j^{g,1}\right\}_{j=1}^{N_g}\right) \tag{3.66}$$

$$0 = \left(\boldsymbol{V^{g,1}}\right) \tag{3.67}$$

$$0 = \left(\Lambda^{\boldsymbol{g,1}}, \Lambda^{g,2}\right) \tag{3.68}$$

spatial grid points inbetween, $l = (\mathrm{g}, 2), \ldots, (\mathrm{g}, N-1)$

$$\frac{\partial Y_i^{g,l}}{\partial t} = \left((\rho v_z)^{g,l}, T^{g,l-1}, T^{g,l}, T^{g,l+1}, Y_i^{g,l-1}, Y_i^{g,l}, Y_i^{g,l+1}\right) \tag{3.69}$$

$$\frac{\partial T^{g,l}}{\partial t} = \left((\rho v_z)^{g,l}, T^{g,l-1}, T^{g,l}, T^{g,l+1}, \left\{Y_j^{g,l-1}, Y_j^{g,l}, Y_j^{g,l+1}\right\}_{j=1}^{N_g}\right) \tag{3.70}$$

$$0 = \left((\rho v_z)^{g,l-1}, (\boldsymbol{\rho v_z})^{\boldsymbol{g,l}}, V^{g,l}, T^{g,l}, \left\{, Y_j^{g,l}\right\}_{j=1}^{N_g}\right) \tag{3.71}$$

$$0 = \left((\rho v_z)^{g,l}, \Lambda^{g,l}, V^{g,l-1}, \boldsymbol{V^{g,l}}, V^{g,l+1}, T^{g,l}, \left\{Y_j^{g,l-1}, Y_j^{g,l}, Y_j^{g,l+1}\right\}_{j=1}^{N_g}\right) \tag{3.72}$$

$$0 = \left(\Lambda^{\boldsymbol{l}}, \Lambda^{l+1}\right) \tag{3.73}$$

last spatial grid point (g, N) (gas-inlet)

$$0=\left(\left\{\boldsymbol{Y}_j^{g,N}\right\}_{j=1}^{N_g}\right) \tag{3.74}$$

$$0=\left(\boldsymbol{T}^{g,N}\right) \tag{3.75}$$

$$0=\left((\boldsymbol{\rho v_z})^{g,N-1},(\rho v_z)^{g,N},V^N,\left\{Y_j^N\right\}_{j=1}^{N_g}\right) \tag{3.76}$$

$$0=\left(\boldsymbol{V}^{g,N}\right) \tag{3.77}$$

$$0=\left((\boldsymbol{\rho v_z})^{g,N},\left\{Y_j^{g,N}\right\}_{j=1}^{N_g}\right) \tag{3.78}$$

In the above equations, the indices for the species are subscript and the indices for the grid point due to spatial discretization are superscript. For instance, $\left\{Y_j^{\mathrm{g},1}\right\}_{j=1}^{N_\mathrm{g}}$ represents the mass fraction of the jth species at the gas-washcoat interface $(\mathrm{g},1)$, $\left\{Y_j^{\mathrm{g},N}\right\}_{j=1}^{N_\mathrm{g}}$ represents the mass fraction of the jth species at the Nth grid point (g, N) (the grid point at the inlet) and $T^{\mathrm{g},l}$ represents the temperature at the lth grid point, and so on.

Dirichlet boundary conditions, which simply specify a certain value for the dependent variables such as $V=0$, are seen as simple constraints that raise the index to one [25]. Since the continuity equation at the inlet is an algebraic constraint, it is differentiated once with respect to time to yield an equation for the radial momentum equation. There is not any explicit boundary condition for Λ. However, the value of Λ at the inlet boundary is determined in a way that all the other boundary conditions are satisfied [25]. The following table show which equation gives rise to an ODE for a certain variable.

Table 3.1: The equations that raise the index to one for a certain variable

Equation number	Index	Variable
(3.63)	0	θ_i
(3.64)	0	$Y_i^{g,1}$
(3.65)	0	$T^{g,1}$
(2.13)	1	$(\rho v_z)^{g,1}$
(3.67)	1	$V^{g,1}$
(3.68)	1	$\Lambda^{g,1}$
(3.69)	0	$Y_i^{g,l}$ for $l = (g, 2), \dots, (g, (N-1))$
(3.70)	0	$T^{g,l}$ for $l = (g, 2), \dots, (g, (N-1))$
(3.71)	1	$(\rho v_z)^{g,l}$ for $l = (g, 2), \dots, (g, (N-2))$ $V^{g,l}$ for $l = (g, (N-1))$
(3.72)	1 1	$V^{g,l}$ for $l = (g, 2), \dots, (g, (N-2))$ $\Lambda_{g,l}$ for $l = (g, (N-1))$
(3.73)	1	Λ^{l} for $l = (g, 2), \dots, g, (N-2)$ $\Lambda^{g,N}$ for $l = (g, (N-1))$
(3.74)	1	$Y_i^{g,N}$
(3.75)	1	$T^{g,N}$
(3.76)	1	$(\rho v_z)^{g,N-1}$
(3.77)	1	$V^{g,N}$
(3.78)	1	$(\rho v_z)^{g,N}$

2) *In case of indirect coupling of RD-approach and DGM surface models with the surrounding flow, the index of the system still remains one. Because, the surface equations consist of only ODEs, therefore their indexes are zero.*

3) *The dependencies of the model equations on different variables regarding the system of equations with the direct coupling of RD-approach with the surrounding flow is given as follows:*

last spatial computational grid point in the washcoat $(wc, N-1)$

$$\partial_t C_i^{wc,N-1} = \left(T^{g,1}, \{C_j^{wc,N-2}, C_j^{wc,N-1}, C_j^{wc,N}\}_{j=1}^{N_g}, \{\theta_k\}_{k=1}^{N_s}\right) \tag{3.79}$$

$$\partial_t \theta_i^{wc,N-1} = \left(T^{g,1}, \{C_j^{wc,N-1}\}_{j=1}^{N_g}, \{\theta_k\}_{k=1}^{N_s}\right) \tag{3.80}$$

spatial grid points $l = (\mathrm{g}, 2), \ldots, (\mathrm{g}, N-1)$

$$\frac{\partial C_i^{wc,l}}{\partial t} = \left(T^{g,1}, \{C_j^{wc,l-1}, C_j^{wc,l}, C_j^{wc,l+1}\}_{j=1}^{N_g}, \{\theta_k\}_{k=1}^{N_s}\right) \tag{3.81}$$

$$\frac{\partial \theta_i^{wc,l}}{\partial t} = \left(T^{g,1}, \{C_j^{wc,l}\}_{j=1}^{N_g}, \{\theta_k\}_{k=1}^{N_s}\right) \tag{3.82}$$

first spatial grid point in the washcoat

$$\frac{\partial C_i^{wc,1}}{\partial t} = \left(T^{g,1}, \left\{C_j^{g,1}, C_j^{wc,1}, C_j^{wc,2}\right\}_{j=1}^{N_g}, \{\theta_k\}_{k=1}^{N_s}\right) \tag{3.83}$$

$$\frac{\partial \theta_i^{wc,1}}{\partial t} = \left(T^{g,1}, \{C_j^{wc,1}\}_{j=1}^{N_g}, \{\theta_k\}_{k=1}^{N_s}\right) \tag{3.84}$$

first spatial grid point in the surrounding flow $(\mathrm{g}, 1)$ (gas-washcoat interface)

$$\frac{\partial Y_i^{g,1}}{\partial t} = \left(T^{g,1}, T^{g,2}, \left\{Y_j^{g,1}, Y_j^{g,2}\right\}_{j=1}^{N_g}, \{C_j^{wc,1}\}_{j=1}^{N_g}\right) \tag{3.85}$$

$$\frac{\partial T^{g,1}}{\partial t} = \left(T^{g,1}, T^{g,2}, \left\{Y_j^{g,1}, Y_j^{g,2}\right\}_{j=1}^{N_g}, \{C_j^{wc,1}\}_{j=1}^{N_g}\right) \tag{3.86}$$

$$0 = \left((\boldsymbol{\rho v_z})^{\boldsymbol{g,1}}, T^{g,1}, \left\{ Y_j^{g,1} \right\}_{j=1}^{N_g} \right) \tag{3.87}$$

$$0 = \left(\boldsymbol{V^{g,1}} \right) \tag{3.88}$$

$$0 = \left(\Lambda^{\boldsymbol{g,1}}, \Lambda^{g,2} \right) \tag{3.89}$$

spatial grid points $l = (\mathrm{g}, 2), \ldots, (\mathrm{g}, N-1)$

$$\frac{\partial Y_i^{g,l}}{\partial t} = \left((\rho v_z)^{g,l}, T^{g,l-1}, T^{g,l}, T^{g,l+1}, Y_i^{g,l-1}, Y_i^{g,l}, Y_i^{g,l+1} \right) \tag{3.90}$$

$$\frac{\partial T^{g,l}}{\partial t} = \left((\rho v_z)^{g,l}, T^{g,l-1}, T^{g,l}, T^{g,l+1}, \left\{ Y_j^{g,l-1}, Y_j^{g,l}, Y_j^{g,l+1} \right\}_{j=1}^{N_g} \right) \tag{3.91}$$

$$0 = \left((\rho v_z)^{g,l-1}, (\boldsymbol{\rho v_z})^{\boldsymbol{g,l}}, V^{g,l}, T^{g,l}, \left\{, Y_j^{g,l} \right\}_{j=1}^{N_g} \right) \tag{3.92}$$

$$0 = \left((\rho v_z)^{g,l}, \Lambda^{g,l-1}, \boldsymbol{V^{g,l}}, V^{g,l+1}, V^{g,l}, T^{g,l}, \left\{ Y_j^{g,l-1}, Y_j^{g,l}, Y_j^{g,l+1} \right\}_{j=1}^{N_g} \right) \tag{3.93}$$

$$0 = \left(\Lambda^{l} \right) \left(\Lambda^{\boldsymbol{l}}, \Lambda^{l+1} \right) \tag{3.94}$$

last spatial grid point (g, N) (gas-inlet)

$$0 = \left(\left\{ \boldsymbol{Y_j^{g,N}} \right\}_{j=1}^{N_g} \right) \tag{3.95}$$

$$0 = \left(\boldsymbol{T^{g,N}} \right) \tag{3.96}$$

$$0 = \left((\rho v_z)^{g,N-1}, (\boldsymbol{\rho v_z})^{\boldsymbol{g,N}}, V^{N}, \left\{ Y_j^{N} \right\}_{j=1}^{N_g} \right) \tag{3.97}$$

$$0 = (V^{g,N}) \tag{3.98}$$

$$0 = \left((\rho v_z)^{g,N}, \left\{Y_j^{g,N}\right\}_{j=1}^{N_g}\right) \tag{3.99}$$

The index of the system remains one, because the surface equations consist only of ODEs.

3.4.3 DETCHEM^{STAG}

The mentioned SFR model, which is explained in sections 3.4.1 and 3.4.2 in detail, is implemented via the new computer code, DETCHEM^{STAG}, which is integrated into DETCHEM software [60]. DETCHEM^{STAG} is the first computational code, which has incorporated mass transport limitations, with different surface models with different complexities, in a porous catalytic layer. The code is validated with experiments for different chemical compositions, reaction mechanisms, temperatures and flow rates as given in the next chapters.

DETCHEM^{STAG} is a useful tool to investigate the interactions between a catalytically active surface and the surrounding flow. Therefore, it can be used to investigate physical and chemical processes in the gas-phase and in the washcoat, and their interactions. Hence, these are discussed in this thesis for certain applications. The code can also be used for the reaction mechanism development purposes. In this respect, it has already been used in [15, 20, 91].

DETCHEM^{STAG} connects to the DETCHEM library, in which some parameters are calculated. These parameters are; mean molar mass ($\overline{M}$), heat capacity (c_{p}), thermal conductivity (λ) and viscosity (μ) of the mixture as well as heat capacity ($c_{\text{p},i}$), molar mass (M_i) and enthalpy (h_i) of each species, averaged diffusion coefficient ($D_{i,\text{M}}$), Knudsen diffusion coefficient ($D_{i,\text{Knud}}$), effective diffusion coefficient ($D_{i,\text{eff}}$), potential gas-phase reaction rate ($\dot{\omega}_i$) and surface reaction rate $\dot{s}_i$.

The index of the model equations, which are used in SFR model, is one therefore a DAE solver LIMEX [86] is used in DETCHEM^{STAG} for the collocation discretization of the DAE system with respect to time. LIMEX solves linearly-implicit differential-algebraic systems of the form,

$$B(t,y) \cdot y'(t) = f(t,y) \quad . \tag{3.100}$$

The solver implements a semi-implicit Euler discretization through,

$$y(t_0 + h) = y(t_0) + h\left[B - hf_y\left(y(t_0)\right)\right]^{-1}\left(fy(t_0)\right) \tag{3.101}$$

in which B is a diagonal matrix. f_y is the Jacobian matrix. It calculates the partial derivatives of the functions with respect to all dependent variables as

$$\begin{pmatrix} \frac{\partial f_1}{\partial y_1} & \frac{\partial f_1}{\partial y_2} & \cdots & \frac{\partial f_1}{\partial y_n} \\ \frac{\partial f_2}{\partial y_1} & \frac{\partial f_2}{\partial y_2} & & \frac{\partial f_2}{\partial y_n} \\ \vdots & \vdots & \ddots & \vdots \\ \frac{\partial f_2}{\partial y_1} & \frac{\partial f_2}{\partial y_1} & \cdots & \frac{\partial f_n}{\partial y_n} \end{pmatrix} \tag{3.102}$$

Discretized form of the SFR equations forms a banded Jacobian. A banded Jacobian can be represented in general as shown in Eq. (3.103).

$$\begin{pmatrix} \frac{\partial f_1}{\partial y_1} & & & & & \\ \frac{\partial f_2}{\partial y_1} & \frac{\partial f_2}{\partial y_2} & \frac{\partial f_2}{\partial y_3} & & & \\ & \frac{\partial f_3}{\partial y_2} & \frac{\partial f_3}{\partial y_3} & \frac{\partial f_3}{\partial y_4} & & \\ & & & & \ddots & \\ & & & \ddots & \ddots & \\ & & & & & \frac{\partial f_{n-1}}{\partial y_n} \\ & & & & \frac{\partial f_n}{\partial y_{n-1}} & \frac{\partial f_n}{\partial y_n} \end{pmatrix} \tag{3.103}$$

The solver enables defining the lower and upper bandwidths in the Jacobian, which reduces the computing time to evaluate the Jacobian. The discretized form of the SFR equations depending on the considered surface models are already explained in section 3.4.2. In this case, the lower and upper bandwidths can be given for different systems of equations as:

- System of equations, which emerge with *∞-approach and η-approach* surface models: $N_g + 4 + \max(N_g + 4, N_s)$
- System of equations with indirect coupling: $N_g + 4$ in the gas-phase, $N_g + N_s$ in the washcoat
- System of equations with direct coupling: due to the reason that there are different number of equations in the gas-phase and in the washcoat, upper and lower bandwidth of the Jacobian cannot be assigned to the LIMEX solver anymore. In this case, LIMEX can solve only the full Jacobian matrix. The consequences and advantages/disadvantages of direct coupling due to large number of reacting species are discussed in the following chapter on CO oxidation results.

Limex has an inbuilt Newton iteration. Newton iteration determines the consistent initial values before the time iteration (at $t = t_0$) for the algebraic and differential variables that satisfies the DAE exactly. The values of the differentiable variables are not changed during the consistent initial value calculation. DETCHEM^{STAG} always uses the Newton iteration option.

4 CO Oxidation on Rh/Al_2O_3

In this chapter, direct oxidation of carbon monoxide (CO) over a porous Rh/Al_2O_3 catalyst is chosen as an example to apply the developed models and computational tool DETCHEM^{STAG}. A recently established SFR is used to provide the experimental data and all necessary information to quantify the characteristics of the catalyst. The main results of this section is published in [20].

4.1 Theoretical background

Catalytic CO oxidation on noble metal surfaces is a simple but important reaction because it produces only gaseous CO_2 as the product, which hardly sticks to metal surfaces, but it still exhibits many of the fundamental steps of a heterogeneous catalytic process [92, 93]. The effect of surface characteristics on reaction kinetics can be investigated at an atomic scale. Therefore, this reaction has been studied extensively in the literature [20, 93-97], regarding the heterogeneous catalysis studies, to understand the relation between the fundamental surface science and practical applications. For instance, CO oxidation is an important reaction for the removal of hazardous CO emission in the automotive exhaust catalyst, in which precious noble metals are used. Furthermore, CO is undesirable in ammonia synthesis and fuel cell power generation systems. Because it reduces the hydrogen productivity, and poison the catalyst in downstream processes. In this case, the undesirable CO content can be removed by using noble metal catalysts. Since the price of the precious noble metals is high, understanding the catalytic CO oxidation at a fundamental level aids optimizing the processes and the catalysts.

It is mostly accepted that CO oxidation on noble metals follows a Langmuir-Hinshelwood mechanism meaning that the reaction occurs between the CO and O adsorbates [20, 98].

4.2 Surface Reaction Mechanism for CO Oxidation over Rh/Al_2O_3

The intrinsic kinetics of the CO oxidation over Rh/Al_2O_3 is taken here from the recent study of Karakaya et al. [99] without any modification. This surface reaction mechanism is a subpart of the kinetics of the water-gas shift reaction over Rh/Al_2O_3 catalysts given by [99]. This direct oxidation of CO involves ten elementary-like surface reaction steps among four surface and three gas-phase species. The reaction rates are modeled by a modified Arrhenius expression as given in Eq. (2.72).

The nominal values of the pre-exponential factors are assumed to be $10^{13} N_A/\Gamma$ (cm^2/mol.s) where N_A is Avagadro's number. The nominal value of 10^{13} is the value calculated from transition state theory ($\mathrm{k_B}T/\mathrm{h}$) with $\mathrm{k_B}$ is being Boltzmann's constant and h is Plank's constant [100].

Exactly the same kinetics of adsorption and desorption of oxygen as well as the reaction of adsorbed oxygen (O(s)) have also been used before to model hydrogen oxidation [15]. The surface reaction kinetics for CO oxidation is given in Table 4.1. The reaction kinetics are thermodynamically consistent at temperatures of 273-1273K.

The rate constants are given in the form of $k=AT^{\beta}$ exp($-E_a/RT$); adsorption kinetics is given in the form of sticking coefficients; the surface site density is $\Gamma=2.72 \times 10^{-9}$ mol cm^{-2}.

Table 4.1: Reaction mechanism for CO oxidation on Rh, taken from [20]

	Reaction	$A^{\dagger}$ (cm,mol,s)	$\beta(-)^{\ddagger}$	Ea (kJ/mol)
R1	O_2+Rh(s)+Rh(s) → O(s)+O(s)	1.000×10^{-2b}	stick. coeff.	
R2	CO_2+Rh(s) → CO_2(s)	4.800×10^{-2b}	stick. coeff.	
R3	CO + Rh(s) → CO(s)	4.971×10^{-1b}	stick. coeff.	
R4	O(s)+O(s) → Rh(s)+Rh(s)+O_2	5.329×10^{22}	-0.137	387.00
R5	CO(s) → CO+Rh(s)	1.300×10^{13}	0.295	134.07-47θCO
R6	CO_2(s) → CO_2+Rh(s)	3.920×10^{11}	0.315	20.51
R7	CO_2(s)+Rh(s) → CO(s)+O(s)	5.752×10^{22}	-0.175	106.49
R8	CO(s)+O(s) → CO_2(s)+Rh(s)	6.183×10^{22}	0.034	129.98
R9	CO(s)+Rh(s) → C(s)+O(s)	6.390×10^{21}	0.000	174.76
R10	C(s)+O(s) → CO(s)+Rh(s)	1.173×10^{22}	0.000	92.14

4.3 Experiment

The experiments of the CO oxidation in SFR was employed by Karakaya [97]. In this section, the catalyst preparation, catalyst characterization and catalytic measurements are explained briefly as it was given in [20].

4.3.1 Catalyst Preparation

The flat stagnation disk was coated with Rh/Al_2O_3 catalyst, where rhodium particles were distributed in a porous Al_2O_3 washcoat. Appropriate amounts of aqueous solution of rhodium (III) nitrate (Umicore) (9 wt.% Rh) and boehmite (AlOOH) (20 % boehmite) were mixed to obtain a 5 wt.% Rh/Al_2O_3 composition. The solution was diluted with water and applied to the disk by the spin-spray technique to ensure a homogeneously distributed catalytic layer on the surface. Coating a flat surface with a well-defined particle size and morphology is essential for the stagnation-flow reactor application [101, 102]. For this purpose a simple laboratory-scale spray apparatus was developed. The stagnation surface was heated to 373 K and held on a rotary support which spins at 1000 rpm. The solution was sprayed by compressed air via a spray gun. The surface was dried at 403 K for 10 min and the procedure was repeated until the desired coating thickness of 100-130 µm is achieved. The coated stagnation disk was then calcined at 973 K in air for 2 h. Prior to the measurements, the surface was oxidized by 5 vol.% O_2 diluted in Ar at 773 K for 2 h. The resulting rhodium oxide phase was reduced by 5 vol.% H_2 diluted in Ar at 773 K for 2 h.

4.3.2 Catalyst Characterization

The coating thickness and the homogeneity of the coating layer were investigated by means of light microscopy (LM: Rechert MEF4A). LM investigations showed that there was a uniform ~100 µm catalyst layer on the supporting disc as shown in Fig. 4.1 [15].

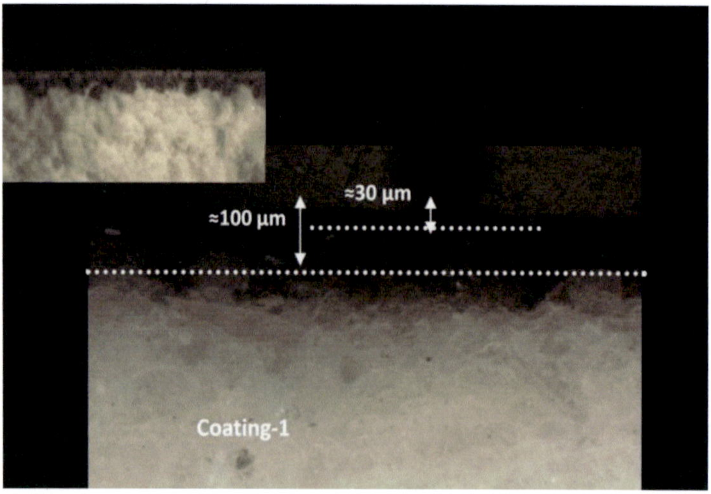

Figure 4.1: LM images of the catalyst, the figure is taken from [98]

For the investigation of nano-scale Rh particles and the washcoat structure, scanning electron microscopy (SEM: Hitachi S570) was applied in combination with energy-dispersive X-ray spectroscopy (EDX) and high resolution transmission electron microscopy (HR-TEM: Philips CM200 FEG). SEM images (Fig. 4.2) indicated a diverse particle size distribution, where as Rh particles of ~100 nm diameter as well as smaller Rh particles of 15-50 nm were also detected in HR-TEM investigations (Fig. 4.3) [15].

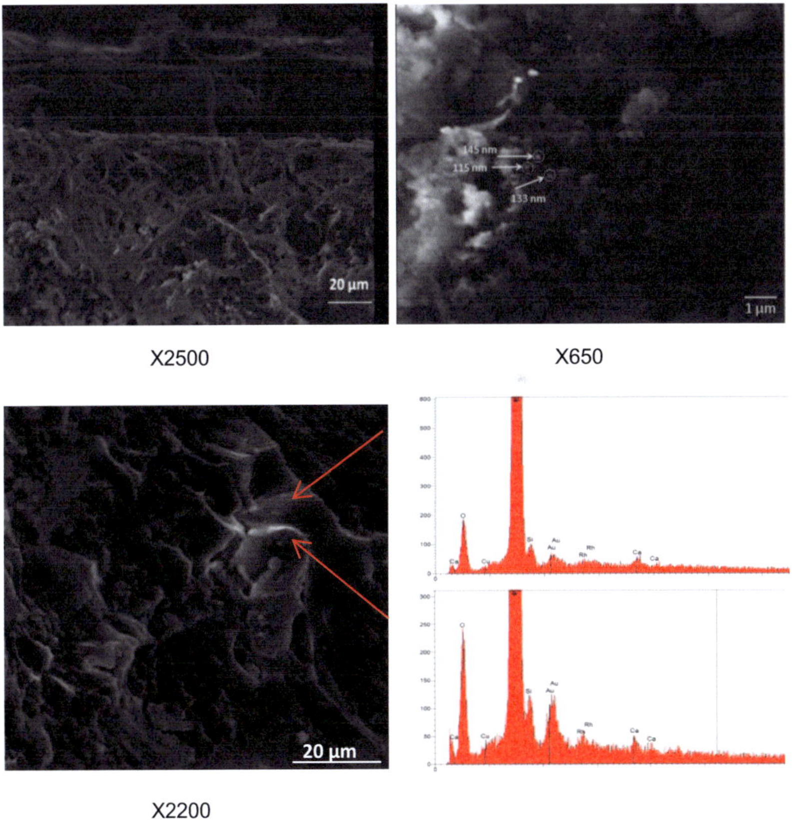

Figure 4.2: SEM-EDX images of the catalyst, the figure is taken from [98]

Metal dispersion was measured by the continuous-flow CO chemisorption technique [103]. The flat stagnation disk was subjected to the chemisorption measurement before the catalytic measurements. The catalytic surface area was calculated to be 0.21 m^2/g based on the CO chemisorption measurements with the assumption of 1:1 adsorption stochiometry between Rh and CO molecules. With this information $F_{cat/geo}$ was calculated to be 30, i.e., the total amount active catalytic surface area equals 30 times the geometrical area of the disk surface.

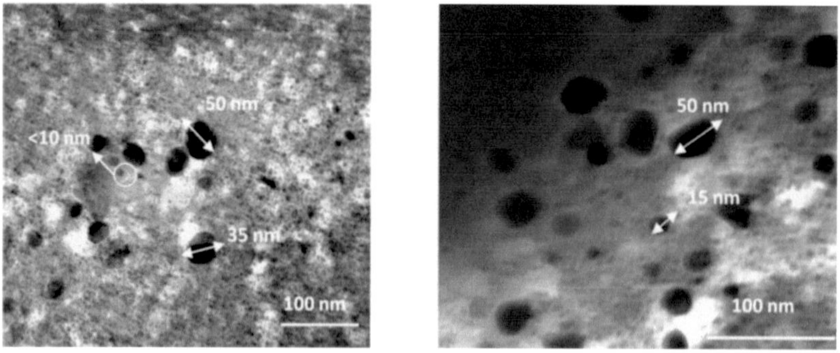

Figure 4.3: TEM images, the figure is taken from [98]

4.3.3 Catalytic Measurements

CO oxidation measurements were carried out in the stagnation-flow reactor at varying CO/O_2 ratios. Ar-diluted gas mixtures were fed to the reactor with a flow rate of 15.5 SLPM (standard liter per minute at 293 K, 1 atm). The calculated flow velocity and working pressure were 51 cm/s and 500 mbar, respectively. The reactor inlet temperature was 313 K. The reaction was studied at steady-state conditions (Table 4.2).

Table 4.2: Stagnation disc temperature and inlet conditions

Case	T_{disc} (K)	T_{inlet} (K)	CO (%vol.)	O_2 (% vol.)	Ar (carrier gas) (% vol.)	Inlet velocity (cm/s)	Reactor pressure (mbar)
1	521	313	2.67	2.23	95.10	51	500
2	673	313	5.67	2.89	91.44	51	500
3	873	313	5.66	2.83	91.51	51	500

The boundary-layer concentration profile of CO, CO_2 and O_2 were measured by using a chemical ionization mass spectrometer (Airsense 500, V&F) with a quadrupole ion trap that. A microprobe sampling technique was used to measure the gas-phase composition in the boundary-layer adjacent to the catalyst surface. Further details on the stagnation-flow reactor and the sampling technique are given elsewhere [15].

4.4 Results and Discussion

4.4.1 Cases Studied

In this work, the experimental stagnation-flow reactor data is used to illustrate the applicability of the developed 1D model. In the experiments, CO oxidation kinetics were investigated at steady-state temperatures of 521 K, 673 K, and 873 K. The reaction conditions are given in Table 4.2. At low temperatures, oxygen-rich conditions were selected to avoid external mass transport limitations and examine the kinetic effects (Case 1). However, for moderate and high temperature regimes (Case 2 and Case 3) the reactions were examined under stoichiometric conditions.

4.4.2 Input data for the numerical simulations

The inlet conditions for the numerical simulations are based on the experimental conditions. Inlet flow velocity is taken as 51 cm/s. This ensures a laminar flow in the reactor and the establishment of the potential flow conditions to apply the model under the given assumptions. The finite gap between the inlet and catalytic surface is 3.9 cm. The surface temperature and inlet mole fractions are given in Table 4.2.

The simulations are performed with all three different models for internal diffusion to analyze the effect of internal mass transfer limitations on the system. The thickness, mean pore diameter, tortuosity and porosity of the washcoat are the parameters that are used in the effectiveness factor approach and the reaction-diffusion equations. The values for these parameters are given in Table 4.3. $F_{cat/geo}$ is taken as 30 according to the chemisorption measurements of Karakaya et al. [15]. The mean pore diameter, which is assumed to be 10 nm, lies in the mesapore range given in literature [27, 104]. CO is chosen as the rate-limiting species for the *η-approach* simulations. *η-approach* simulations are also performed with considering O_2 as the rate-limiting species.

Table 4.3: The parameters used in the effectiveness factor approach and reaction-diffusion equations

Thickness of the washcoat (μm)	Mean pore diameter (nm)	$F_{cat/geo}$	Porosity (%)	Tortuosity
100	10	30	60	3

4.4.3 Boundary-layer Thickness

Boundary layer thickness is investigated first with the effectiveness factor approach for Case 1, Case 2 and Case 3 with the inputs given in Table 4.2 and Table 4.3. Concerning the boundary layer formed on top of the catalytic disc, the scaled radial velocity profile for Case 1 reveals that the viscous boundary-layer stands adjacent to the surface (x<0.3 cm) (Fig. 4.4), and the axial velocity monotonically decreases from its maximum at the inlet to zero on the surface. The thermal and species boundary-layer thicknesses are approximately 0.45 cm (Fig. 4.4).

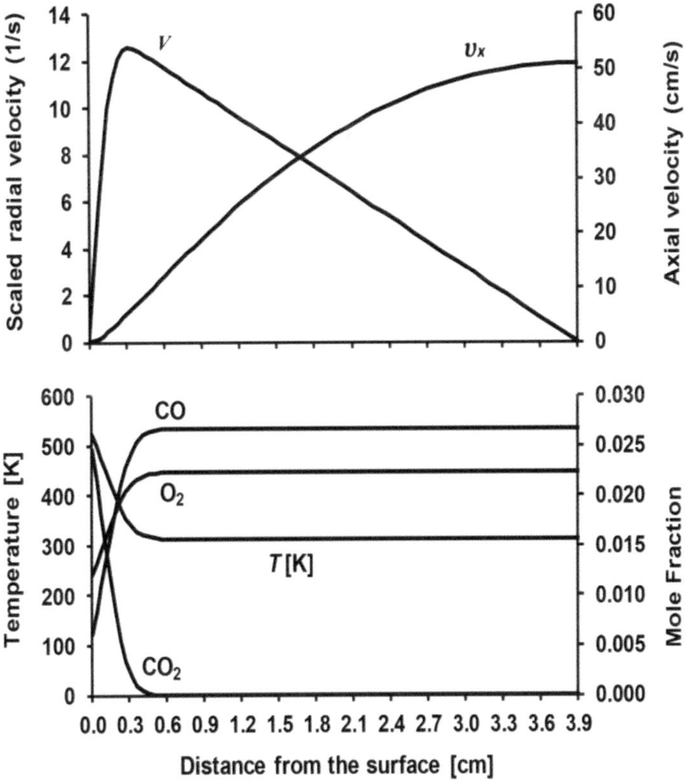

Figure 4.4: Simulation results of velocity profiles by using the *η-approach* (*V:* scaled radial velocity, v_z: axial velocity), temperature and species boundary thicknesses at 521K, the grids are generated by using the simple aspect ratio (number of grid points: 40, aspect-ratio: 1.03)

Axial velocity and scaled radial velocity profiles for Case 2 and Case 3 are found to be similar as in Case 1 as they are given in Fig. 4.5 and Fig. 4.6, respectively . Due to the increased surface temperature, thermal and species boundary thicknesses expand to 0.5 and 0.6 cm for Case 2 and Case 3, respectively.

The boundary thicknesses in all three cases (Fig. 4.4, Fig. 4.5 and Fig. 4.6) are found to be small relative to the lateral extent of the reactor fulfilling the pre-condition for the one-dimensional finite-gap SFR model.

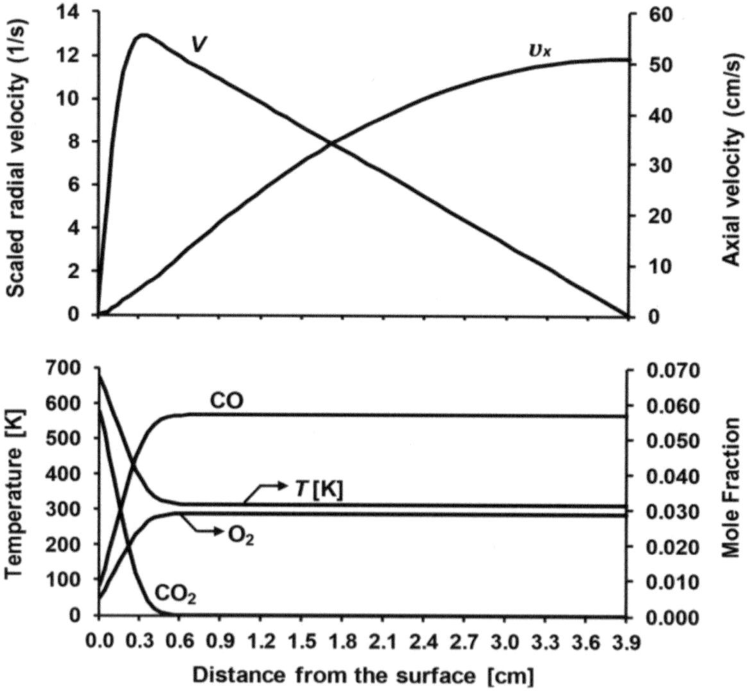

Figure 4.5: Simulation results of velocity profiles by using the *η-approach* (*V:* scaled radial velocity, v_z: axial velocity), temperature and species boundary thicknesses at 673K, the grids are generated by using the simple aspect ratio (number of grid points: 40, aspect-ratio: 1.03)

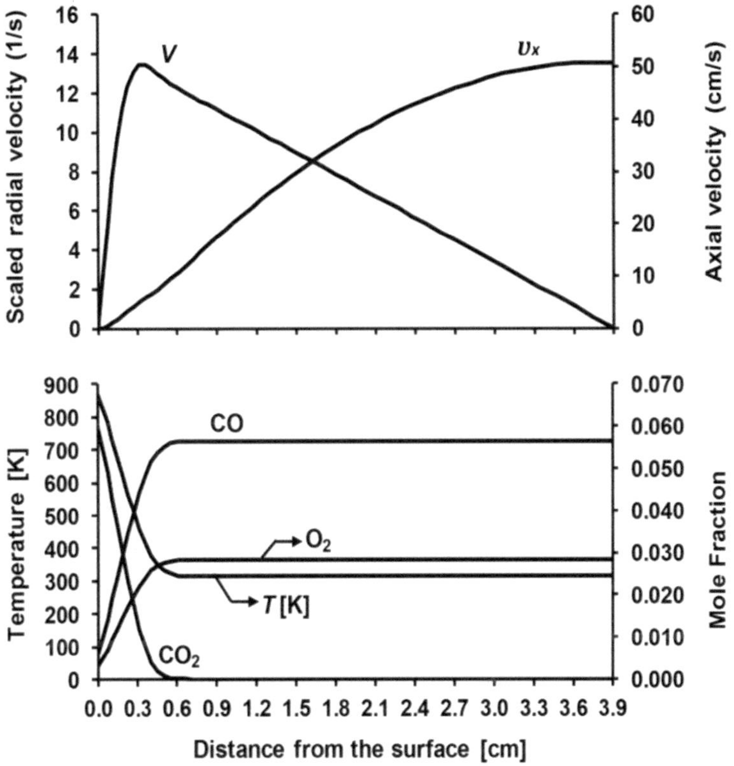

Figure 4.6: Simulation results of velocity profiles by using the *η-approach* (*V:* scaled radial velocity, v_z: axial velocity), temperature and species boundary thicknesses at 873K, the grids are generated by using the simple aspect ratio (number of grid points: 40, aspect-ratio: 1.03)

4.4.4 Fluid Compressibility

It is mentioned in chapter 3 that the constant pressure formulations of the 1D SFR model are derived through further simplifications on compressible Navier-Stokes equations. In this section, fluid compressibility is discussed shortly. Gas flows can be approximated as incompressible if the change of density is less than 5% [105]. For nonisothermal and reacting flow the density changes significantly as a function of temperature and pressure through the equation of state (Eq. (2.30)). In SFR cases, the density changes greatly in the boundary layer in the SFR due to high temperature and species concentration gradients as illustrated for Case 3 in Fig. 4.7. Density change reaches %63 on the surface relative to the inlet for Case 3. However, the flow is laminar and the pressure is assumed to be constant due to very small pressure variations compared to the mean thermodynamic pressure. Therefore, the stagnation flow solution shows the characteristics of incompressible flow.

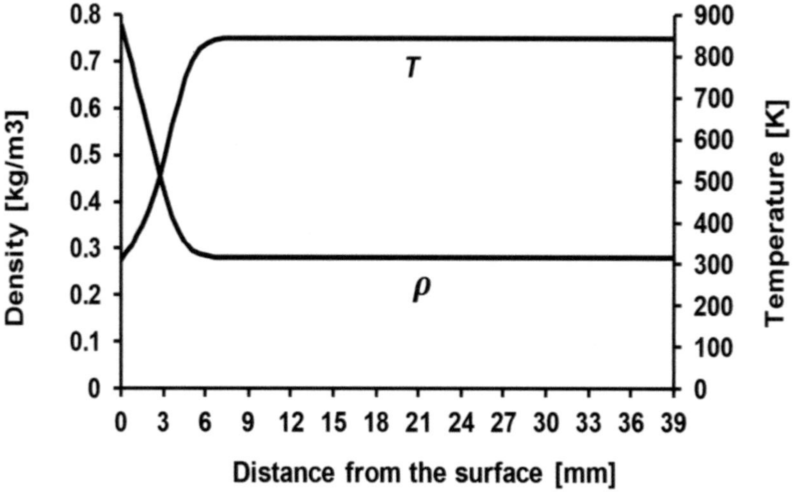

Figure 4.7: The change of density in the SFR with respect to temperature (*η-approach* is used for the simulations, the grids are generated by using the simple aspect ratio, number of grid points: 40, aspect-ratio: 1.03), indirect coupling scheme is applied

4.4.5 Species Profiles

The reaction is already active at 521K (Fig. 4.8), but total consumption of the reactants is not achieved in the experiment. However, the *∞-approach* predicts complete consumption of CO at the surface, i.e., it strongly overpredicts the overall reaction rate. Simulations with the *η-approach* and *RD-approach* models predict the slow overall reaction rate of the experiments. The slight deviation for the O_2 consumption might be due to sampling inaccuracies in the experiment. The *RD-approach* predicts the species profiles inside the porous washcoat, for the first case, as given in Fig. 4.9. Species are consumed or produced just within the first 7−7.5 µm of the washcoat. This can be attributed to the fact that surface reactions are very fast even at this low temperature. The rate-limiting process is already internal diffusion. *η-approach* yields Thiele modulus Φ = 27.4 and effectiveness factor η = 0.04, respectively, confirming the strong diffusion limitation.

In the second case (T = 673K), CO and O_2 concentration at the surface decrease by 82% and 71%, respectively, relative to the inlet conditions (Fig. 4.10). *∞-approach* predicts total consumption for both reactants. Simulations with the *RD-approach* surface model estimate results close to the experiments for the consumption of reactants and production of CO_2. There is a relatively good agreement between the experiment and the simulation results with the *η-approach* surface model, as well. Species profiles inside the washcoat (predicted with *RD-approach*) are similar to Case 1; but the reaction layer decreases from 7.5 to 6.5 µm (Fig. 4.11). For this condition, the dimensionless Φ and η are calculated as 53.7 and 0.02, respectively.

In the last case (T = 873K), CO and O_2 concentration at the surface decrease by 84% and 79%, respectively, relative to the inlet conditions (Fig. 4.12). CO_2 formation has its maximum value, since reaction rate reaches its maximum. *∞-approach* underpredicts consumption of reactants, and overpredicts formation of CO_2. Simulation with the *RD-approach* surface model reproduces the experimental data. There is also a relatively good agreement again between the experiments and the simulation with the *η-approach* surface model. At this temperature, reactions are even faster, resulting in large concentration gradients within the first 5.5-6 µm in the washcoat (Fig. 4.13). The *Φ* and *η* are 91.7 and 0.01, respectively.

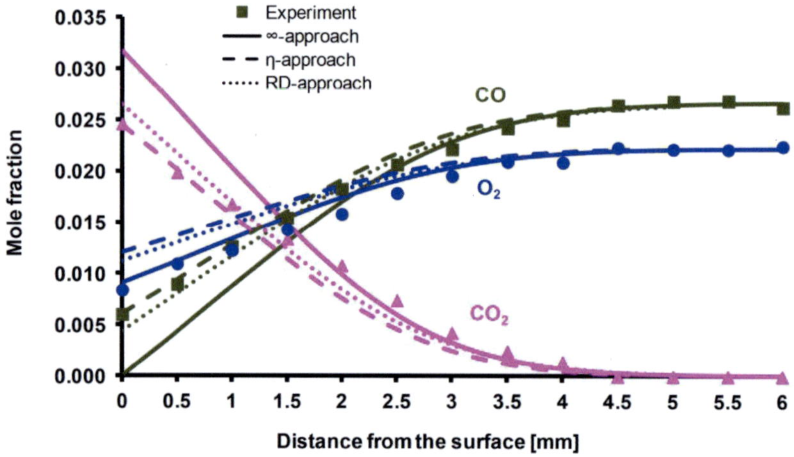

Figure 4.8: Comparison of the experimental and simulation results for the species profiles in catalytic oxidation of CO at 521 K, simple aspect ratio is used for grid generation, indirect coupling method is used for the *RD-approach* simulations

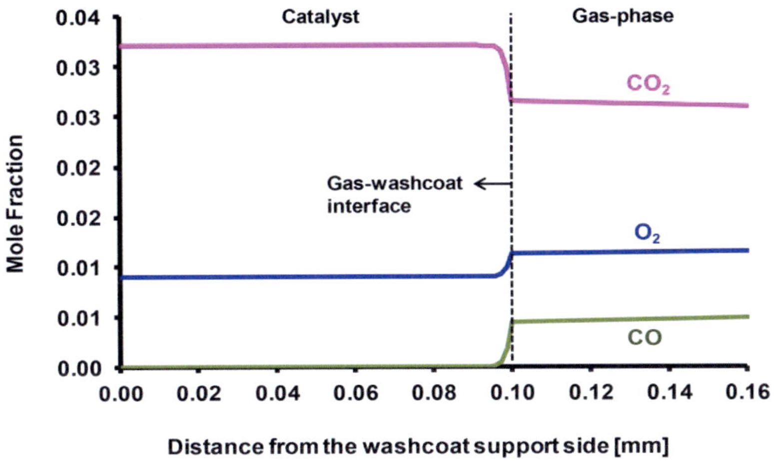

Figure 4.9: Species mole fractions inside the porous washcoat layer at 521K (*RD-approach*), simple aspect ratio is used for grid generation, indirect coupling method is used for the simulations

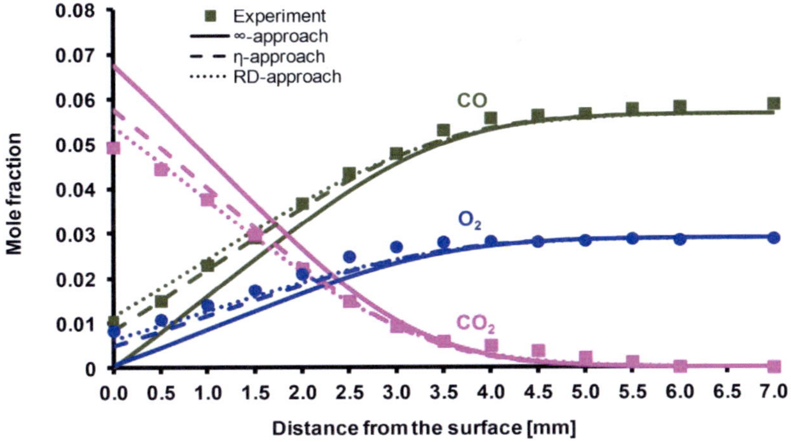

Figure 4.10: Comparison of the experimental and simulation results for the species profiles in catalytic oxidation of CO at 673 K, simple aspect ratio is used for grid generation, indirect coupling method is used for the simulations

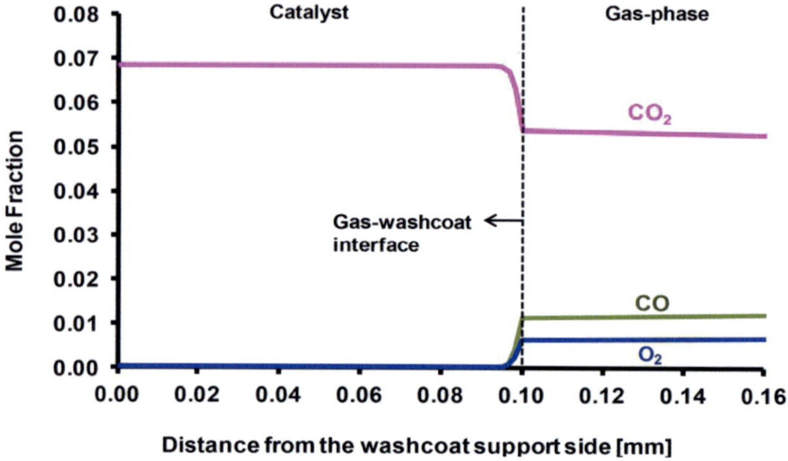

Figure 4.11: Species mole fractions inside the porous washcoat layer at 673K (*RD-approach*), simple aspect ratio is used for grid generation, indirect coupling method is used for the simulations

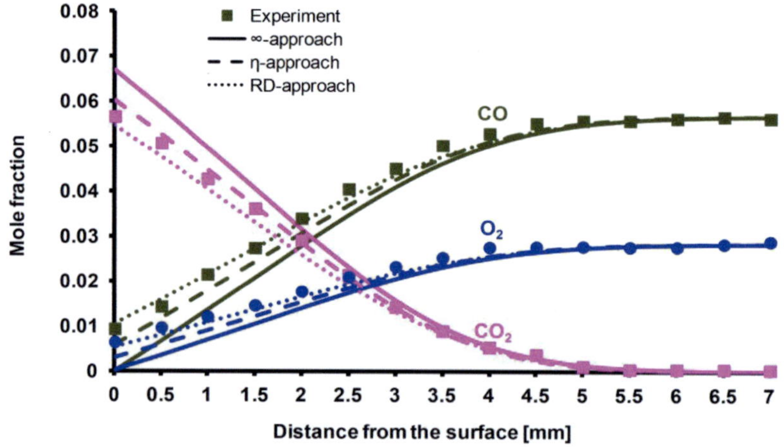

Figure 4.12: Comparison of the experimental and simulation results for the species profiles in catalytic oxidation of CO at 873 K, simple aspect ratio is used for grid generation, indirect coupling method is used for the simulations

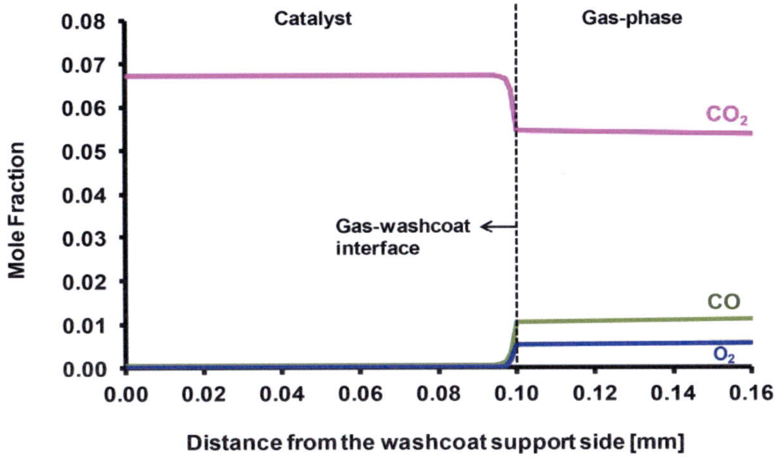

Figure 4.13: Species mole fractions inside the porous washcoat layer at 873K, simple aspect ratio is used for grid generation, indirect coupling method is used for simulations

Finally, *η-approach* simulations are performed for considering O_2 as the rate-limiting species. In this situation, *η-approach* overpredicts the consumption of CO and formation of CO_2 for the lean Case 1 (521 K) (Figure 4.14a). Considering O_2 or CO as the rate-limiting species gives the same results with the stoichiometric experiments for Case 2 (Figure 4.14b) and for case 3, respectively (Figure 4.15).

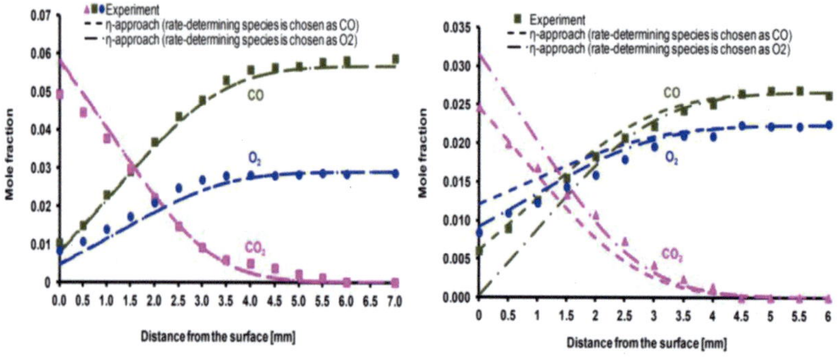

Figure 4.14: Comparing η-approach simulations by considering CO and O_2 as the rate-limiting species at (a) 521K and (b) 673K, simple aspect ratio is used for grid generation, indirect coupling method is used for the simulations

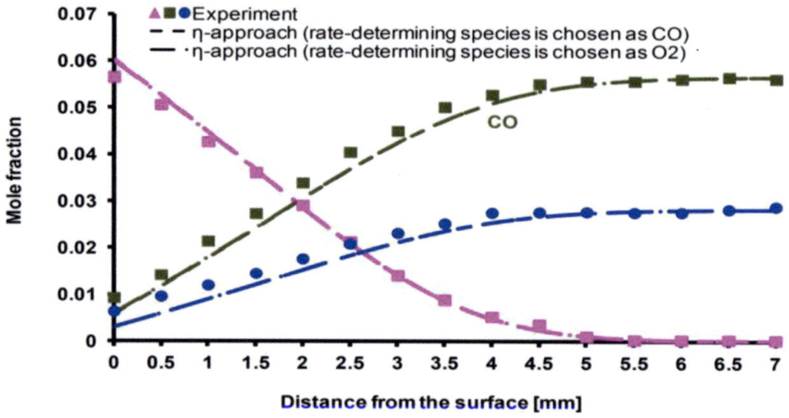

Figure 4.15: Comparing *η-approach* simulations by considering CO and O_2 as the rate-limiting species at 873K, simple aspect ratio is used for grid generation, indirect coupling method is used for the simulations

4.5 The effect of finer mesh near the gas-washcoat interface

In the previous simulation results only the aspect ratio is used for a finer mesh resolution near the gas-washcoat interface. Number of grid points in the gas-phase and in the washcoat, and the aspect ratios are given in Table 4.4.

If there is only equidistant mesh with the same number of grid points in the gas-phase and in the washcoat (40 and 30 respectively), the results deviate from the experiments at 873 K as shown in Fig. 4.16. If the number of the grid points in the gas-phase and in the washcoat is increased to 80 and 50, respectively, the results get closer to the experiments. However, more grid points result in more computational cost, especially when reaction-diffusion equations are solved.

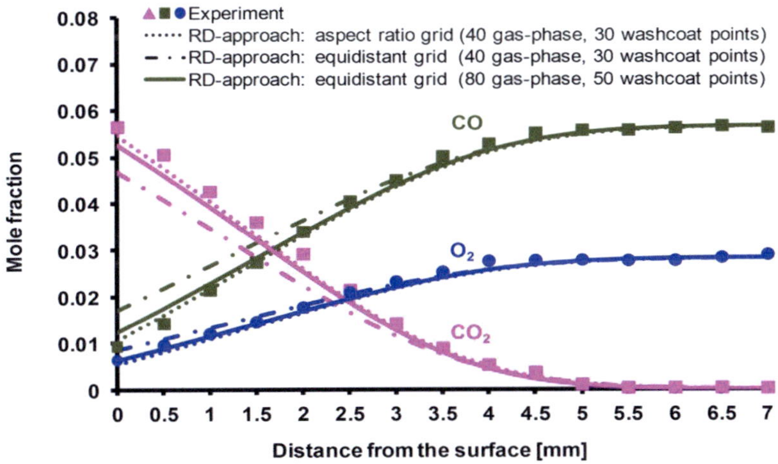

Figure 4.16: Comparison of the species profiles in the gas-phase in catalytic oxidation of CO at 873 K with equidistant and aspect ratio grid generation, indirect coupling method is applied

Table 4.4: Number of grid points and aspect ratios in the gas-phase and in the washcoat

	Number of grid points	Aspect ratio
Gas-phase	40	1.03
Washcoat	30	1.06

Another possibility is using an adaptive gridding. When the adaptive gridding is applied, the Limex code [86] does not enable to change the size of the equation system during a simulation. Therefore, when the DETCHEM^{STAG} code decides to insert a new grid point, it quits the simulation, adds new grid point, and restarts the simulation with the new number of grid-points in case of *∞-approach* and *η-approach*. The computational expense of this solution procedure is considerably low. However, in contrast, this solution procedure can be very time consuming when the *RD-approach* is used. Therefore, the following approach is followed when the adaptive gridding is applied to the *RD-approach*: initially the number of grid points in the gas-phase is determined by implementing the adaptive gridding by using the *η-approach.* Then *RD-approach* simulation is started with the actual grid-points in the gas-phase (obtained from adaptive gridding by using the *η-approach*). Number of grid-points in the gas-phase does not change anymore. However, adaptive gridding is applied then for the grid points in the washcoat throughout the simulation.

In the following figure, adaptive gridding results and locations of the grid points along the axial distance is shown. *RD-approach* simulations are initialized with 10 points in the gas-phase and 6 points in the washcoat. The simulation ended with 38 mesh points in the gas-phase and 29 mesh points in the washcoat as shown in Fig. 4.17 and Fig. 4.18.

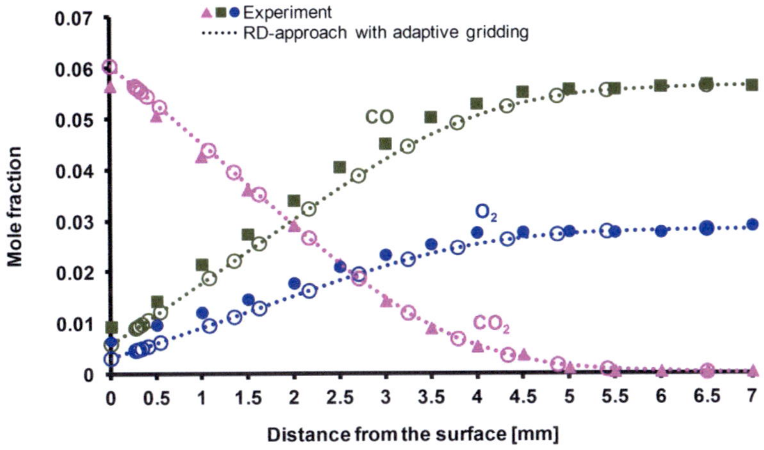

Figure 4.17: Species profiles in the gas-phase in catalytic oxidation of CO at 873 K: the grids are generated with adaptive grid method, indirect coupling method is used for the simulations

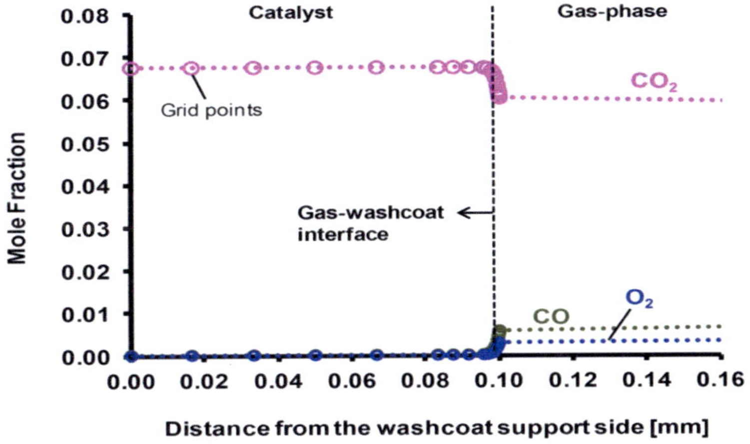

Figure 4.18: Species profiles in the washcoat in catalytic oxidation of CO at 873 K: the grids are generated with adaptive grid method by using the RD-approach, indirect coupling method is used for the simulations

4.6 The effect of direct and indirect coupling of washcoat equations with the surrounding gas-phase

Direct and indirect coupling of the washcoat and surrounding gas-phase equations, and lower and upper bandwidths are already explained in section 3.4.2 and section 3.4.3. In this section, the results, advantages and disadvantages of both coupling method is discussed.

The results of both coupling method are discussed based on the conditions of Case 3, with different examples with different grid points and aspect ratios. These examples are given in Table 4.5. It is seen in Table 4.6 that the difference in species mole fractions, which are obtained from direct and indirect coupling method simulations, are less than 1.E-08 for all examples. Therefore, it can be concluded that both method gives nearly the same results.

Direct coupling offers faster simulations. Indirect coupling costs more computational time. In case of direct coupling, the Jacobian matrix becomes unstructured due to different number of equations in the gas-phase and in the washcoat. Therefore, upper and lower bandwidth of the Jacobian cannot be assigned to the LIMEX solver anymore, and LIMEX can solve only the full Jacobian matrix. The solution of the all coupled nonlinear equations requires accurate grid resolution throughout the simulation for faster convergence. In this case, adaptive gridding method generates the grids automatically throughout the simulation, which helps to avoid numerical instabilities that might occur in simulations with fixed number of grid points in direct coupling case.

The objective of this study is producing the steady-state results by using a transient iteration strategy, not producing transient results. That is the reason of nearly identical results of direct and indirect coupling. However, they will create different results in transient studies. Because, in case of indirect coupling, the surface models are iterated separately until they reach steady state as it is explained in section 3.4.2. For transient cases such as catalytic ignition, gas-phase and washcoat equations should be directly coupled to capture the transient dynamics of the catalytic surface and the surrounding flow. On the other hand, it should be mentioned that the numerical solution strategy will also alter for transient simulations [25], which is not considered in this study.

Table 4.5: Examples for the comparison of direct and indirect coupling methods

	Gas-phase grid points / aspect ratio	Washcoat grid points/ aspect ratio
Example 1 (for Case 3)	30 / 1.03	30 / 1.06
Example 2 (for Case 3)	25 / 1.05	25 / 1.1
Example 3 (for Case 3)	25 / 1.05	25 / 1.2

Table 4.6: Simulation results of examples given in Table 4.4

	CO mole fraction	CO2 mole fraction	O2 mole fraction
Example 1 direct coupling	0.01011137839711	0.05505338594011	0.005127314714693
Example 1 indirect coupling	0.01011137831442	0.05505338619953	0.005127314668192
Example 2 direct coupling	0.009267928551584	0.05603838433472	0.004709698238651
Example 2 indirect coupling	0.009267928285070	0.05603838475846	0.004709698107053
Example 3 direct coupling	0.006518984432414	0.05929519597597	0.003342369861624
Example 3 indirect coupling	0.006518984044120	0.05929519660660	0.003342369604418

4.7 Comparing DETCHEM^{STAG} simulations with the CHEMKIN SPIN code results

In this section, two different simulations are performed exemplarily to compare the results of DETCHEM^{STAG} and CHEMKIN SPIN code. The boundary conditions are taken from Case 2 and Case 3. The simulation results for Case 2 and Case 3 were given in [98] by using CHEMKIN SPIN code. Since SPIN code does not account for the internal mass transfer limitations directly, Karakaya multiplied all the surface reactions by a presumed coefficient to obtain a so-called effective $F_{cat/geo}$ value as given below [97],

$$\left(F_{cat/geo}\right)_{effective} = \eta_i \cdot \left(F_{cat/geo}\right)_{experimental} \quad . \tag{4.1}$$

Karakaya [98] has multiplied all the surface rates with 0.6 for Case 2 and Case 3. Therefore, the same value is taken for DETCHEM^{STAG} simulations. In addition, adaptive grid refinement is used.

It is seen in Fig. 4.19 and Fig. 4.20 that both codes give almost identical results for Case 2 and Case3, respectively.

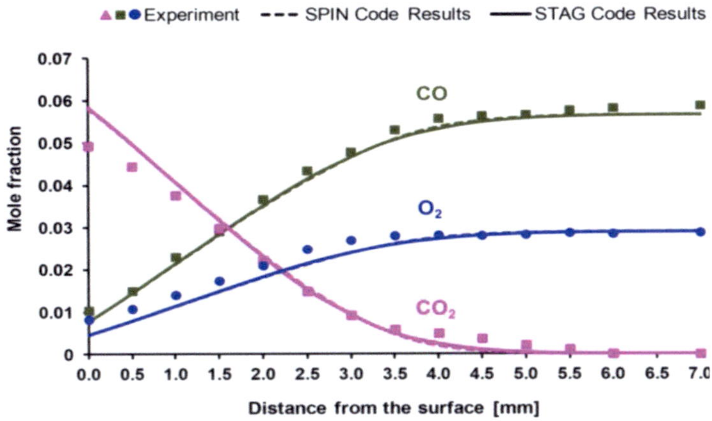

Figure 4.19: Comparing DETCHEM^{STAG} and CHEMKIN SPIN code results for Case 2, all surface reactions are multiplied with the coefficient representing the $(F_{cat/geo})_{effective}$, the grids are generated by using adaptive grid refinement in both codes, CHEMKIN SPIN code results are taken from [97]

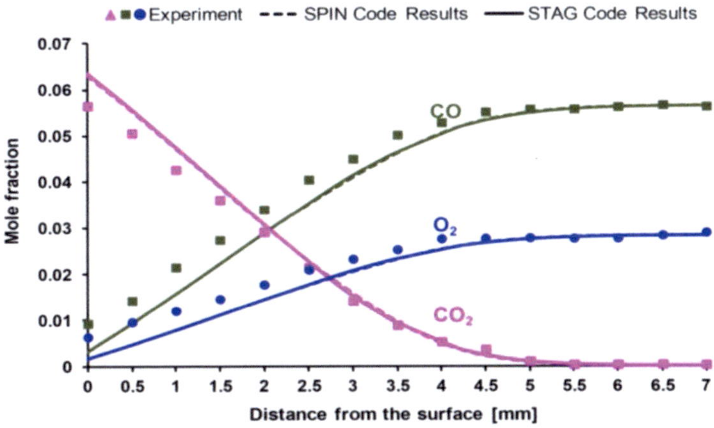

Figure 4.20: Comparing DETCHEMSTAG and CHEMKIN SPIN code results for Case 3, all surface reactions are multiplied with the coefficient representing the $(F_{cat/geo})_{effective}$ the grids are generated by using adaptive grid refinement in both codes, CHEMKIN SPIN code results are taken from [97]

4.8 Conclusions

The one-dimensional mathematical model and computer code, DETCHEM^{STAG}, was applied to investigate direct oxidation of CO over a thick Rh/Al_2O_3 catalyst in a SFR. For this purpose, a recently developed surface reaction mechanism [99] was used for the direct oxidation of CO. Experimental measurements were carried out to evaluate the numerical model and by doing so also the CO oxidation part of the surface reaction kinetics.

Due to the high sticking coefficient of CO on Rh, the reaction rate is very high, even at moderate temperatures, which implies that internal and external mass transfer may play a role in the interpretation of overall measured reaction rates. Indeed, simulations with the *∞-approach* (instantaneous diffusion) were unable to make accurate predictions of the measured species profiles. The overall reaction rate and therefore species profiles were strongly influenced by internal mass transport limitations requiring adequate models. Both models for finite diffusion used in this study can account for this effect. Actually, simulations with the *RD-approach* resolving the spatial profiles inside the washcoat predict the measured species profiles well. The much simpler *η-approach (Thiele modulus)* yields good agreement with the experiments for all the cases studied, when CO was chosen as the rate-limiting species. Since CO is a simple reaction mechanism, there was also a sufficiently good agreement between the simulations using the *η-approach* and the experimentally measured profiles [20].

Direct and indirect coupling of the gas-phase and washcoat equations yielded almost identical results. However, it becomes more difficult for the solver to converge to the results in direct coupling with high number of grid points. DETCHEM^{STAG} and CHEMKIN SPIN code results were compared exemplarily for two cases based on multiplying all the surface reaction rates with the so-called $\left(F_{\text{cat/geo}}\right)_{\text{effective}}$ coefficient. Both codes yielded very close results for the simulated cases.

The new computational code, DETCHEM^{STAG} is applied to more complex systems such as partial oxidation and reforming of hydrocarbons in the next chapters. For those systems with more complex reaction networks, it is expected that simplifications of the effectiveness factor approach will be much more relevant and more sophisticated models are needed such as the *RD-approach* presented here.

It is shown that finer mesh resolution near the external catalyst surface predicts the experiments better than equidistant grid resolution (with the same number of grid points) due to high temperature and species concentration gradients.

5 Water-Gas-Shift Reaction on Rh/Al_2O_3

In this chapter, water-gas-shift (WGS) and reverse water-gas-shift (RWGS) reactions are numerically investigated in stagnation flow on a porous Rh/Al_2O_3 catalyst. The importance of internal mass transfer limitations is already manifested in the previous chapter. Therefore, internal mass transfer resistances are accounted here with the *η-approach* and *RD-approach*. Furthermore, the effect of the convective flow inside the washcoat is investigated with the dusty-gas model (*DGM*). The effect of external mass transfer limitations is investigated based on the Damköhler number. The numerically predicted species profiles in the external boundary layer are compared with recently measured profiles [99]. The influence of flow rates, pressure and washcoat thickness on CO consumption is also examined in this chapter. It is discussed how the mean pore diameter, porosity and tortuosity in the washcoat affects internal mass transfer limitations and CO consumption. Finally, fundamental findings are applied for a commercial WGS catalyst with industrially relevant inlet mole fractions. The main results of this section is published in [91].

5.1 Theoretical Background

The reversible water-gas shift (WGS) reaction Eq. (5.1) is used in many industrial applications. It is one of the most crucial reactions, which affects the product selectivity, in syngas production by total and partial oxidation, steam and dry reforming of hydrocarbons [91, 99, 106-108].

$$CO + H_2O \rightleftharpoons CO_2 + H_2 \quad \Delta H_r = -41.1\ kJ/mol \tag{5.1}$$

Recently, noble metal catalysts have been investigated as the promising next-generation WGS catalysts [109, 110]. In addition, they facilitate the design and development of small scale fuel cell applications such as on-board fuel processors for small scale power vehicles or portable fuel cell system for powering electronic devices as a replacement for batteries [106].

Microreactors, such as monolithic beds, offer a suitable ambient for noble metal catalysts. In monolithic reactors, the active catalyst material is adhered, possibly in a porous layer called washcoat, to the inner wall of the channels. In this case, microchannel reactors with rhodium catalysts offer high conversion, enhanced heat and mass transfer, safe control, high surface area, low pressure drop and short residence time (10ms or less) [111, 112]. In addition, they are slightly prone to carbon-deposition, and stable even at extreme, cyclic conditions without loss of activity [113].

Aforementioned microreactors for the WGS applications exhibit a complex interaction between the catalytically active surface and the surrounding flow field. Understanding the physical and chemical steps of a heterogeneous catalytic process at a fundamental level aids optimizing the process and the catalyst. Microkinetic models are incorporated into computational fluid dynamics (CFD) codes to model the catalytic reactors, and validate them in an operating range relevant to industrial applications. In this respect, there have been studies to understand the kinetics and the mass transfer phenomenon in microreactors regarding the WGS applications [61, 99, 109, 110, 114-116]. Some studies indicate that external and internal mass transfer limitations are negligible [110, 115], whereas others indicate that internal mass transfer limitations are important but external mass transfer limitations are negligible [61]. In this case, more studies are needed to give more insight to physical and chemical processes in microreactors regarding the WGS applications.

5.2 Surface Reaction Mechanism

In this chapter, the recently developed multi-step surface reaction mechanism for WGS and RWGS reactions over Rh/Al_2O_3 catalyst from Karakaya et al. [99] is used without any modification. In this surface reaction mechanism, it is assumed that all the species adsorb on the active metal, i.e., the alumina support does not function as an active site. The thermodynamically consistent mechanism consists of 30 reactions among five gas-phase and eight surface species. The surface reaction mechanism is given in Appendix B. In this study, gas-phase reactions are neglected, because they are unlikely to occur at the considered temperature range.

5.3 Results and Discussion

5.3.1 Cases Studied

In this section, the conditions of our recent stagnation flow experiments of WGS and RWGS over Rh/Al2O3 are used [99]. The WGS reactions were carried out at 873, 1008 and 1073 K with a molar steam/carbon ratio of 1.1. The RWGS was studied at 873 and 973 K with a molar CO2/H2 ratio of 1. The initial conditions of the WGS and RWGS cases studied in this chapter are summarized in Table 5.1.

Table 5.1: Reaction conditions for the considered WGS and RWGS cases

Reaction	Temperature (K)	H_2 (% vol.)	CO (% vol.)	H_2O (% vol.)	CO_2 (% vol.)	Ar(carrier gas) (% vol.)
WGS Case 1	873	-	4.75	5.18	-	90.07
Case 2	1008	-	4.75	5.18	-	90.07
Case 3	1073	-	4.75	5.18	-	90.07
RWGS Case 4	873	5.20	-	-	5.20	89.6
Case 5	973	5.20	-	-	5.20	89.6

5.3.2 Input Data for the Numerical Simulations

The inlet conditions of Case 1-Case 5 are based on experimental conditions. Thus, inlet temperature is taken as 423 K and 313 K for WGS and RWGS cases, respectively. The inlet velocity and reactor pressure are 74 cm/s and 500 mbar, respectively. The finite gap between the inlet and catalytic surface is 3.9 cm.

CO and CO_2 are chosen as the rate limiting species for *η-approach* simulations in WGS and RWGS cases, respectively. The parameters that are used in surface models for Case 1-Case 5 follow: thickness of the washcoat (100 µm), mean pore diameter (10 nm), porosity (40%) and tortuosity (8). $F_{\mathrm{cat/geo}}$ was calculated in [99] as 30. Therefore, the same $F_{\mathrm{cat/geo}}$ value is used in the simulations.

The effect of pressure, flow rates, and washcoat thickness on the CO consumption is discussed with the inlet mole fractions and surface temperature of Case 1. Similarly, the influence of mean pore diameter, porosity and tortuosity is discussed with the inlet conditions of Case 1.

5.3.3 WGS Results

The experimental and simulation results for Case 1 are given in Fig. 5.1. According to the experiments, consumption of the reactants is low at 873 K. As a consequence, production of H_2 and CO_2 are also low. Species boundary layer is around 5 mm. The simulations with the *η-approach*, *RD-approach* and *DGM* surface models show relatively good agreement with the experiments. *RD-approach* and *DGM* simulations give an insight to understand the internal mass transfer limitations inside the washcoat.

Table 5.2: The pressure difference in the washcoat and Damköhler number for WGS cases

	Case 1	Case 2	Case 3
Pressure difference (Pa)	103	94	90
Damköhler Number (Da)	1.58	2.27	2.60

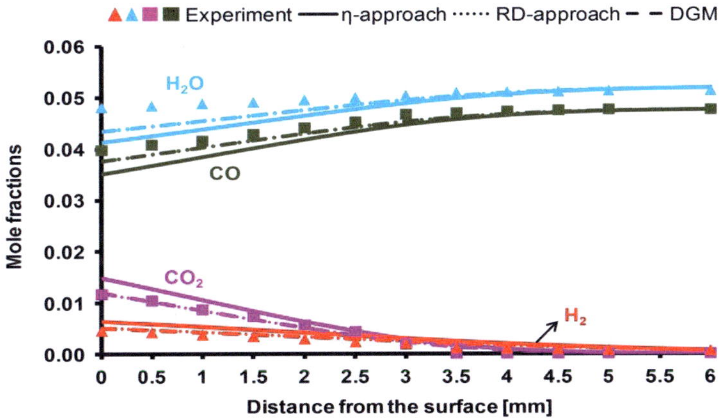

Figure 5.1: Experimental and simulation results for the species profiles in WGS at 873 K, simple aspect ratio and adaptive gridding are used for grid generation (for DGM only aspect ratio is used), indirect coupling method is used for the *RD-approach* and *DGM* simulations

Fig. 5.2 reveals that the reaction layer is 31 µm relative to the external catalyst surface. Surface reactions are fast and internal mass transfer limitations are observed. *η-approach* yields Thiele Modulus $\Phi = 17.9$ and effectiveness factor $\eta = 0.06$, respectively, confirming the strong diffusion limitation. The *DGM* simulation yields that the pressure difference between the gas-washcoat interface and the washcoat support side is low for Case 1 (Table 5.2), which means that the species transport inside the washcoat due to the pressure-driven convective flow is negligible. As a result, the *DGM* simulation yields identical species profiles with the *RD-approach* simulation (Fig.5.1 and Fig.5.2).

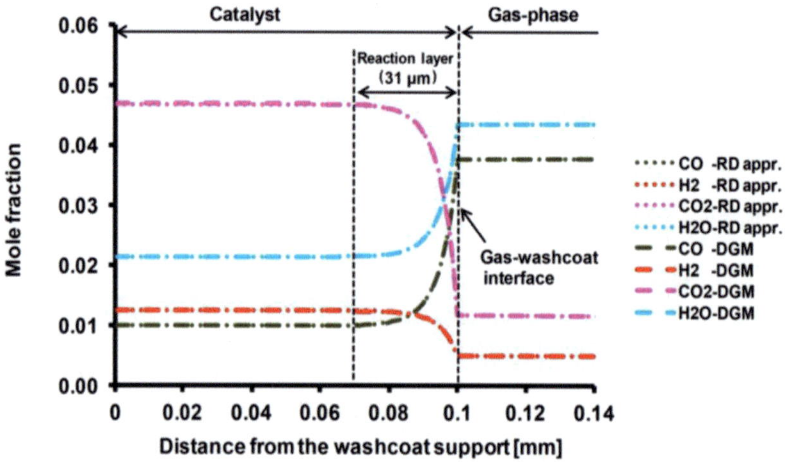

Figure 5.2: Species mole fractions inside the porous washcoat layer in WGS at 873 K, simple aspect ratio and adaptive gridding are used for grid generation (for DGM only aspect ratio is used), indirect coupling method is used for the *RD-approach* and *DGM* simulation

The experimental and simulation results for Case 2 are given in Fig. 5.3. According to the experiments, WGS activity increases due to increased surface temperature. The *η-approach, RD-approach* and *DGM* shows a relatively good agreement with the experiments again. The low pressure difference inside the washcoat (Table 5.2) is again the reason for identical species profiles from the *RD-approach* and *DGM* simulations. The thickness of the reaction layer inside the washcoat decreases to 20 µm, in comparison to Case 1, due to faster surface reactions (Fig. 5.4). Internal mass transfer limitations are prominent on the system. *η-approach* yields $\Phi = 23.8$ and $\eta = 0.04$, respectively.

In case 3, species boundary layer increases to 6 mm due to increased surface temperature (Fig. 5.5). *η-approach*, *RD-approach* and *DGM* simulations all predict the experiments well. According to the *RD-approach* and *DGM* simulations, reaction layer decreases to 18 µm (Fig. 5.6). Internal mass transfer limitations are prominent in this case, as well. *η-approach* yields $\Phi = 26.4$ and $\eta = 0.04$, respectively. The pressure difference in the washcoat is low again (90 Pa).

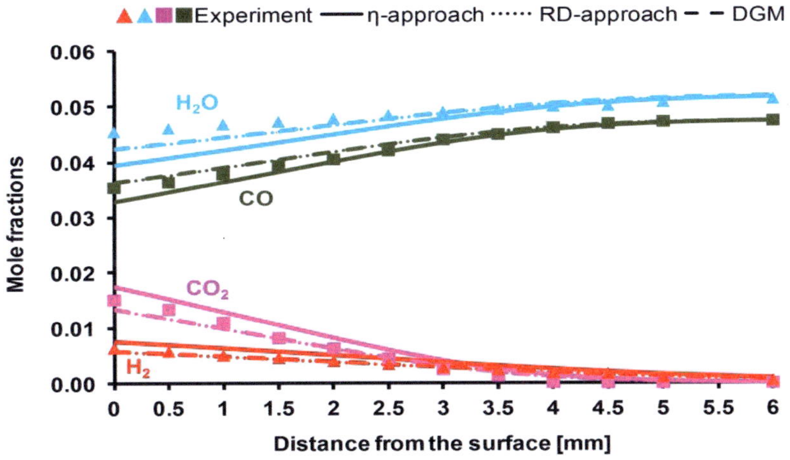

Figure 5.3: Experimental and simulation results for the species profiles in WGS at 1008 K, simple aspect ratio and adaptive gridding are used for grid generation (for DGM only aspect ratio is used), indirect coupling method is used for the *RD-approach* and *DGM* simulations

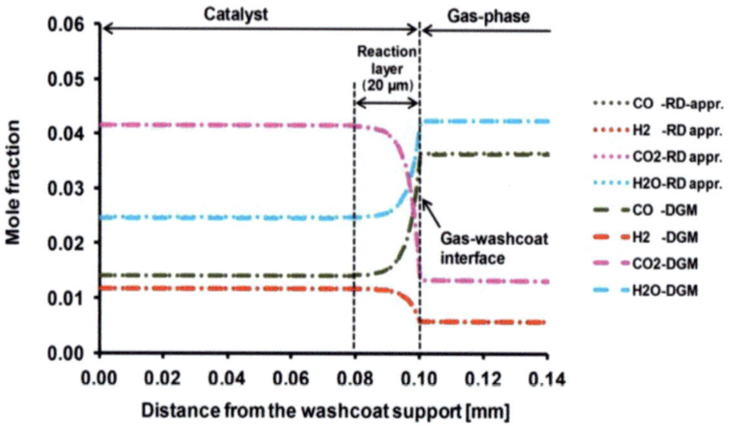

Figure 5.4: Species mole fractions inside the porous washcoat layer in WGS at 1008 K, simple aspect ratio and adaptive gridding are used for grid generation (for DGM only aspect ratio is used), indirect coupling method is used for the *RD-approach* and *DGM* simulations

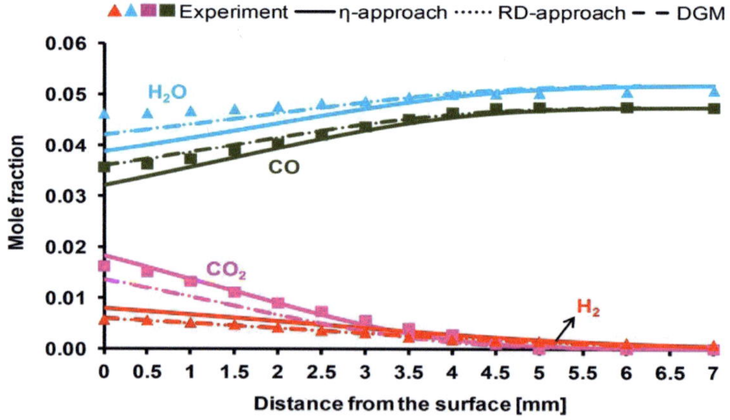

Figure 5.5: Experimental and simulation results for the species profiles in WGS at 1073 K, simple aspect ratio and adaptive gridding are used for grid generation (for DGM only aspect ratio is used), indirect coupling method is used for the *RD-approach* and *DGM* simulations

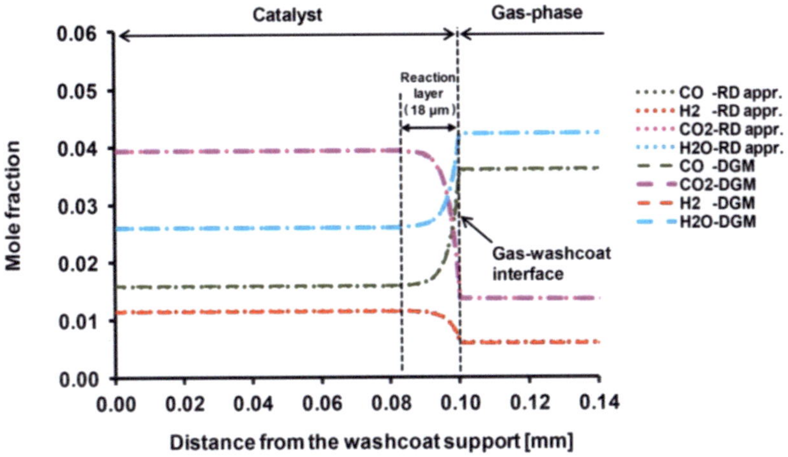

Figure 5.6: Species mole fractions inside the porous washcoat layer in WGS at 1073 K, simple aspect ratio and adaptive gridding are used for grid generation (for DGM only aspect ratio is used), indirect coupling method is used for the *RD-approach* and *DGM* simulations

The effect of external mass transfer limitations on WGS cases are discussed based on the Damköhler (Da) number. The dimensionless Da number relates the reaction rate to the transport phenomena in the system and it can be calculated from a relation between Reynolds (Re), Schmidt (Sc) and Sherwood (Sh) numbers and observed reaction rate (r_{obs}). Re number in the finite gap is first calculated as [8]

$$Re = \frac{\rho v_z L}{\mu} \quad . \tag{5.2}$$

Sc number is calculated as [8]

$$Sc = \frac{\nu}{D_{i,M}} \quad . \tag{5.3}$$

Sh number for a flow over a flat plate can be calculated now based on Re and Sc numbers [117]

$$Sh = 0.66Re^{1/2}Sc^{1/3} \quad . \tag{5.4}$$

The relation between the Sh number and mass transfer coefficient (h_m) is given as [8]

$$Sh = \frac{h_m L}{D_{i,M}} \quad . \tag{5.5}$$

Observed reaction rate is calculated then as [118]

$$r_{obs} = h_m\left(c_i^b - c_i^s\right) \tag{5.6}$$

where c_i^b and c_i^s are the concentration of species i in the bulk gas-phase and stagnation surface, respectively. Finally, Da number can be calculated as

$$Da = \frac{r_{obs}/h_m}{c_i^b - r_{obs}/h_m} \quad . \tag{5.7}$$

If Da is greater than 3, external mass transfer limitation becomes important in the system [119]. The calculated Da numbers for Case 1, Case 2 and Case 3 (Table 5.2) indicate that external mass transfer limitations can be neglected for all the three cases. Since external mass transfer limitations are negligible for the studied WGS cases, the low CO and H_2O consumption can be associated with the internal mass transport limitations due to the thick washcoat layer.

5.3.4 RWGS Results

Case 4 considers the RWGS reaction at 873 K. In this case, RWGS activity is quite low (Fig. 5.7). The CO level is 0.52 vol.% at the surface. The species boundary layer in the gas-phase is around 4 mm. The simulations with all the surface models predict the experiments well. The low pressure difference in the washcoat for Case 4 (Table 5.3) results in identical species profiles from the *RD-approach* and *DGM* simulations. Internal mass transfer limitations are important according to the *RD-approach* and *DGM* simulations. The reactants are consumed and the products are formed within the first 32 µm of the washcoat (Fig. 5.8). *η-approach* yields $\Phi = 9.41$ and $\eta = 0.11$, respectively, confirming the strong diffusion limitation.

In Case 5, the surface temperature is increased to 973 K. However, there is not a significant change in the RWGS activity in comparison to Case 4 (Fig. 5.9). The simulations predict the experiments again. Surface reactions are fast and internal mass transfer limitations are prominent on the system. The whole reaction layer is around 26 µm (Fig. 5.10). *η-approach* yields $\Phi = 12.0$ and $\eta = 0.08$, respectively.

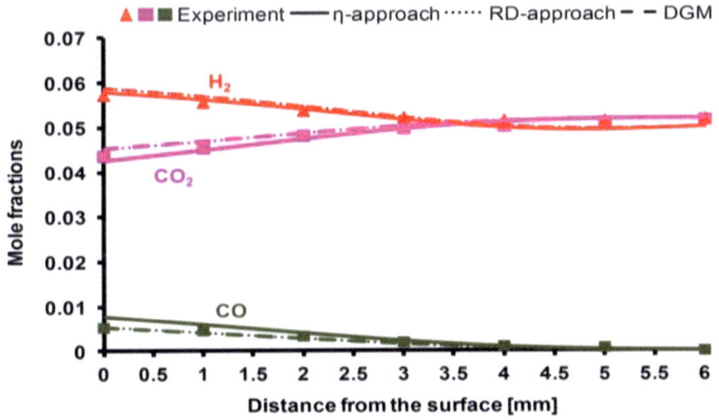

Figure 5.7: Experimental and simulation results for the species profiles in RWGS at 873 K, simple aspect ratio and adaptive gridding are used for grid generation (for DGM only aspect ratio is used), indirect coupling method is used for the simulations

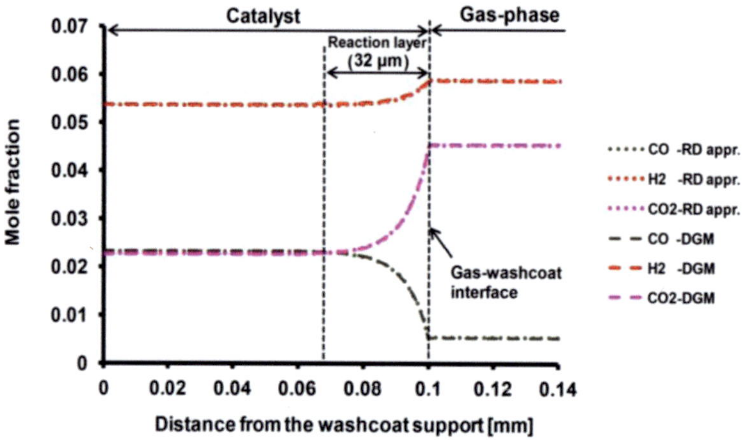

Figure 5.8: Species mole fractions inside the porous washcoat layer in RWGS at 873 K, simple aspect ratio and adaptive gridding are used for grid generation (for DGM only aspect ratio is used), indirect coupling method is used for the *RD-approach* and *DGM* simulations

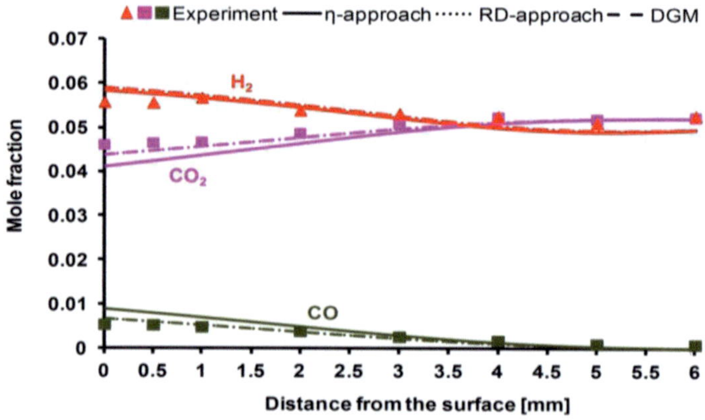

Figure 5.9: Experimental and simulation results for the species profiles in RWGS at 973 K, simple aspect ratio and adaptive gridding are used for grid generation (for DGM only aspect ratio is used), indirect coupling method is used for the *RD-approach* and *DGM* simulations

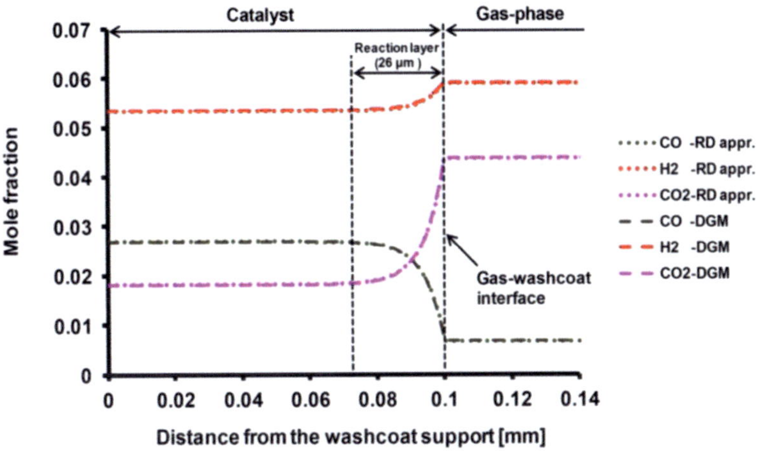

Figure 5.10: Species mole fractions inside the porous washcoat layer in RWGS at 973 K, simple aspect ratio and adaptive gridding are used for grid generation (for DGM only aspect ratio is used), indirect coupling method is used for the simulations

The calculated *Da* number for Case 4 and Case 5 (Table 5.3) indicates that the external mass transfer limitations are negligible for the studied RWGS cases.

Table 5.3: The pressure difference in the washcoat and Damköhler number for RWGS cases

	Case 4	Case 5
Pressure difference (Pa)	66	81
Damköhler Number (Da)	2.28	2.49

5.3.5 The Effect of Pressure, Flow Rates and Washcoat Thickness on the CO Consumption in WGS Reactions

In this section, the effect of the pressure and flow rates on the CO consumption in WGS reaction is investigated. The simulations are initially performed with varying pressures from 0.5 to 3 bar, and varying inlet velocity from 0.2 to 0.9 m/s. The simulation results (Fig. 5.11) show that the mole fraction of CO on the surface decreases with the increasing pressure and decreasing inlet flow velocity. These results are based on two reasons: 1) When the reactor pressure is increased, the number of collisions of the reactants also increases, 2) When the inlet velocity is decreased, the residence time for the reactants increases [120]. This expected observation can be considered in practical WGS catalyst implementations especially for high CO conversion purposes (purification of CO from syngas).

In the subsequent simulations, the effect of the washcoat thickness on CO consumption is investigated over a wide range of temperature range. The pressure and flow rate are chosen as 3 bar and 0.2 m/s, respectively. It is seen in Fig. 5.12 that the maximum CO consumption is obtained around 550 °C similar to observations of CO consumption in WGS on Rh/ceria catalysts [115]. The effect of the washcoat thickness on CO consumption varies with the temperature. At low temperatures until 500 °C, the washcoat thickness does not have any effect because the reactivity is already low. The similar trend is seen at higher temperatures as well (above 800 °C). Between 500-800 °C, in which the chemical reactivity is higher, the washcoat thickness effects the CO consumption. Therefore, increasing the washcoat thickness decreases the consumption of CO. This result is due to existing internal mass transfer limitations. If the same amount of catalyst is put into thinner washcoat layer, the species have to diffuse through shorter distances rather than longer distances, where internal mass transfer limitations exist.

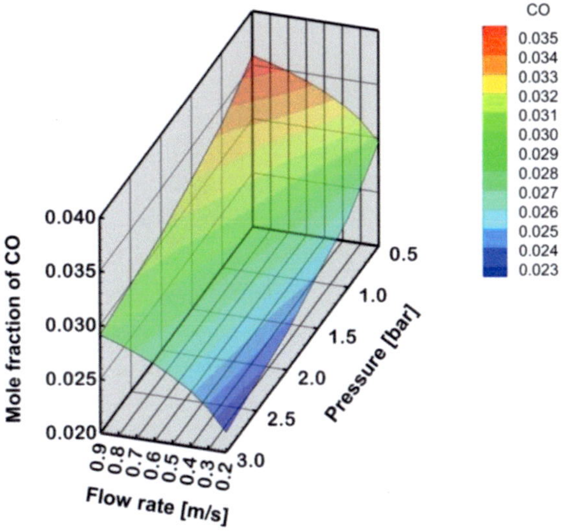

Figure 5.11: Change of CO mole fraction on the surface with respect to pressure and inlet flow velocity, simulations are performed with the *η-approach*

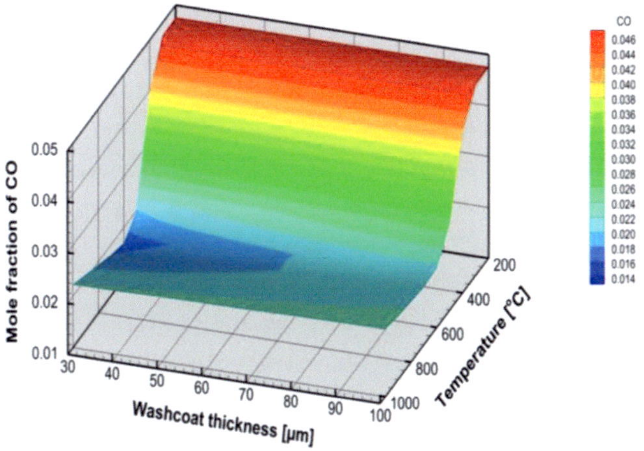

Figure 5.12: Change of CO mole fraction on the surface with respect to washcoat thickness and surface temperature, simulations are performed with *η-approach*

5.3.6 Impact of Mean Pore Diameter, Tortuosity, and Porosity

The effects of mean pore diameter, tortuosity and porosity on internal mass transfer limitations and CO consumption are discussed now for a thick catalyst layer (Catalyst A: 100 µm) and a relatively thinner catalyst layer (Catalyst B: 40 µm). The effect of pore sizes is investigated for micro, meso and macropores. The list of different pore sizes, which are used in the simulations, is given in Table 5.4.

Table 5.4: Investigated pore sizes

	Micropore	Mesopore	Macropore
Mean pore diameter (nm)	1	2.5, 5, 7.5, 10, 25, 50	75, 100, 250, 500, 750, 1000

It is seen in Fig. 5.13 and in Fig. 5.14 that the lowest CO consumption is obtained in the micropore regime (at 1nm) both for Catalyst A and Catalyst B. In this regime, the pore size is very low for species to diffuse easily even in the relatively thin catalyst layer (Catalyst A). Therefore, internal mass transfer limitations are very high. Effectiveness factor is obtained 0.015 and 0.030 for Catalyst A and Catalyst B, respectively (Fig. 5.15 and in Fig. 5.16). The mole fraction of CO on the surface decreases in the mesopore range considerably for both catalysts (Fig. 5.13 and Fig. 5.14), because the species can diffuse easier through bigger pores. Therefore, mass transport limitations decrease. Effectiveness factor reaches 0.135 at 50 nm pore diameter for Catalyst A (Fig. 5.15). It reaches 0.24 at the same pore size for Catalyst B (Fig. 5.16). Increasing the mean pore diameter within the macropore regime continues increasing the CO consumption for both catalysts (Fig. 5.13 and Fig. 5.14). However, increasing the mean pore diameter more than 250 nm does not affect the CO consumption considerably, because the species can already be transported efficiently in the washcoat. At 250 nm, effectiveness factor already reaches 0.30 and 0.52 for Catalyst A and Catalyst B, respectively. The results indicate that CO consumption and effectiveness factor are higher for Catalyst B at all conditions.

In the subsequent simulations, the effect of tortuosity and porosity on internal mass transfer limitations is investigated for Catalyst A and Catalyst B. It is seen in Fig. 5.17 and Fig. 5.18 that decreasing tortuosity and increasing porosity decreases internal mass transfer limitations for both catalysts. When the porosity is increased, pore interconnections are increased. Therefore, species can diffuse easier. When the tortuosity factor is decreased, alternate routes for diffusion become possible, which results in increasing species fluxes [120]. At a mean pore diameter of 10 nm, tortuosity 8 and porosity 0.3, effectiveness factor reaches 0.05 and 0.08 for Catalyst A and Catalyst B, respectively. There is high internal mass transfer limitation in both catalysts. At a mean pore diameter of 10 nm, tortuosity 2 and porosity 0.7, effectiveness factor reaches 0.17 and 0.32 for Catalyst A and Catalyst B, respectively, resulting in lower internal mass transfer limitations.

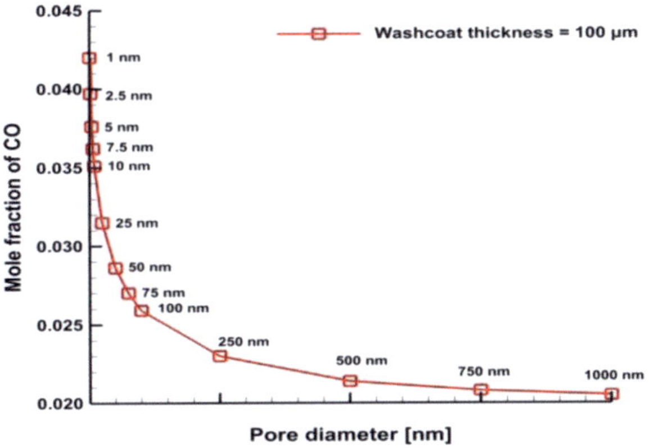

Figure 5.13: The effect of mean pore diameter on CO consumption at 100 μm washcoat thickness

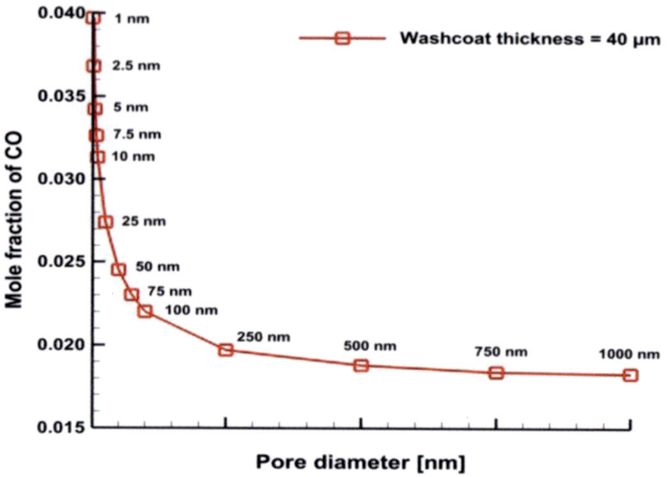

Figure 5.14: The effect of mean pore diameter on CO consumption at 40 µm washcoat thickness

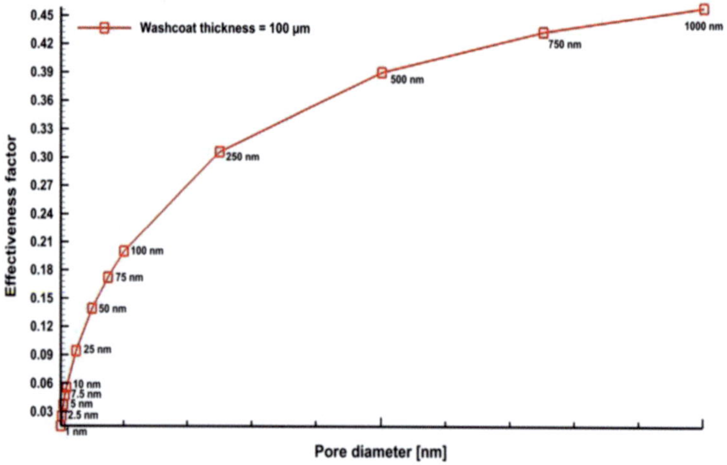

Figure 5.15: The effect of mean pore diameter on internal mass transfer limitations at 100 µm washcoat thickness

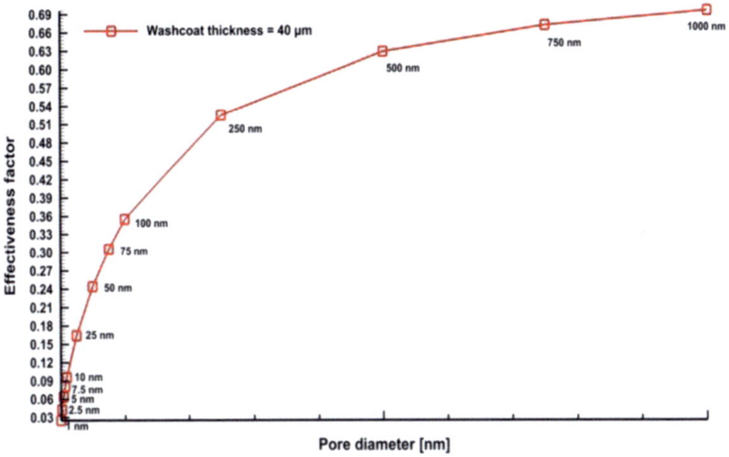

Figure 5.16: The effect of mean pore diameter on internal mass transfer limitations at 40 µm washcoat thickness

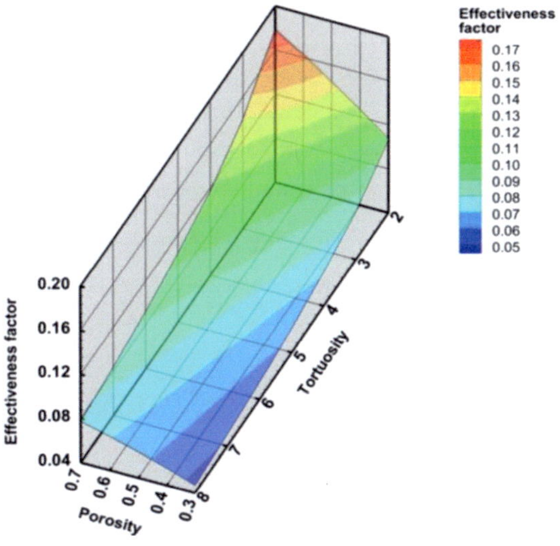

Figure 5.17: The effect of tortuosity and porosity on internal mass transfer limitations at 100 µm washcoat thickness

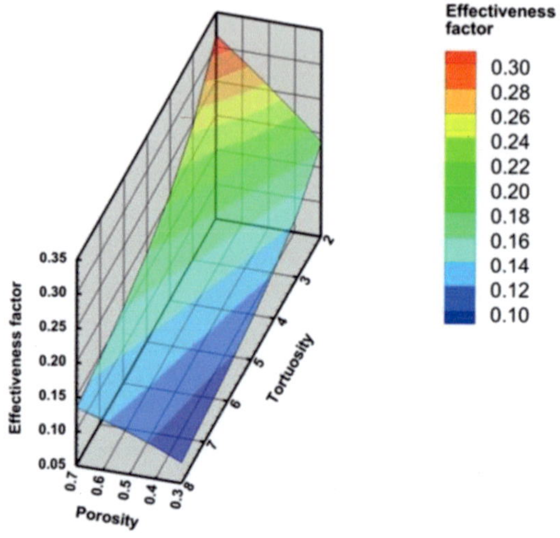

Figure 5.18: The effect of tortuosity and porosity on internal mass transfer limitations at 40 µm washcoat thickness

It can be concluded here that internal mass transfer limitations can be decreased significantly with the decreasing inlet flow velocity, increasing reactor pressure, thinner washcoat layer, higher washcoat mean pore diameter, higher washcoat porosity and lower washcoat tortuosity.

Accordingly, the following test case (Table 5.5) is simulated in SFR to obtain very low internal mass transfer limitations and high CO consumption.

Table 5.5: Test case for obtaining very low internal mass transfer limitations and high CO consumption

Inlet mole fractions	4.75% CO, 5.18% H2O, 90.07% Ar
Inlet flow velocity (m/s)	0.2
Reactor pressure (bar)	3
$F_{\text{cat/geo}}$	30
Mean Pore Diameter (nm)	100
Porosity	0.6
Tortuosity	2
Washcoat Thickness (μm)	40

The simulations with this test case results in effectiveness factor and CO mole fraction at the surface as 0.93 and 0.013, respectively. As expected, internal mass transfer limitations are decreased, and CO consumption is increased significantly.

5.3.7 Applications of findings in monolithic WGS Reactors

The fundamental findings, which were obtained through SFR investigations, are applied now for monolithic WGS reactors. In this respect, single channel of a honeycomb catalyst is simulated for WGS applications. The length and diameter of the channel are 10 and 1 mm, respectively [99, 122]. The inlet molar composition is 32% H_2, 10% CO, 23% H_2O, 8% CO2 and 27% N2, which represents industrially relevant conditions.

Three different test conditions are considered for reactor pressure, inlet flow velocity, washcoat thickness, washcoat mean pore diameter, porosity and tortuosity as summarized in Table 5.6. DETCHEM^{PLUG} code [60] is used to simulate the single channel.

Table 5.6: Different test conditions for a single channel of a WGS catalyst

	Test-Case 1	Test-Case 2	Test-Case 3
Inlet flow velocity (m/s)	0.74	0.2	0.2
Reactor pressure (bar)	1	3	3
Reactor Temperature (K) (isothermal conditions in the channel)	823	823	823
$F_{\text{cat/geo}}$	30	30	30
Mean Pore Diameter	10	10	100
Porosity	0.4	0.4	0.6
Tortuosity	8	8	2
Washcoat Thickness (μm)	100	100	40

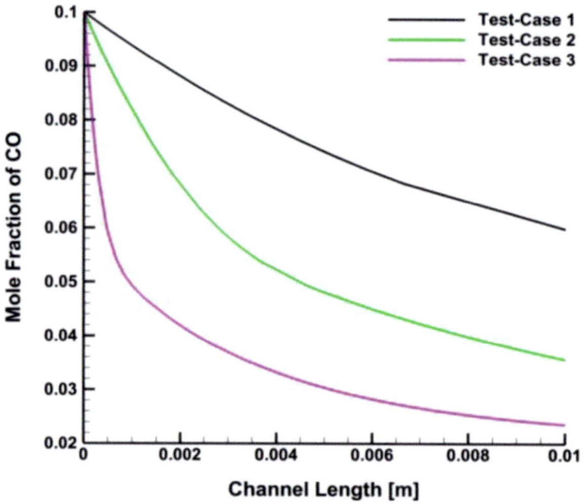

Figure 5.19: Mole fraction of CO in the channel at different conditions, the simulations are performed with DETCHEMPLUG code

It is seen in Fig.19 that maximum CO at the channel outlet is obtained in Test-Case 1. Because, the inlet flow velocity is high, reactor pressure is low and internal mass transfer limitations are prominent due to the catalyst conditions as discussed before. In Test-Case 2, CO at the channel outlet decreases significantly in comparison to Case 1, due to decreased inlet flow velocity and increased reactor pressure. In Test-Case 3, the effect of internal mass transfer limitations is decreased by changing the properties of the catalyst in comparison to Test-Case 2. Consequently, minimum CO at the channel outlet is obtained in Test-Case 3.

5.3.8 Grid Generation

In the preceding WGS simulations with SFR model, grids are generated with the simple aspect ratio and adaptive gridding. The number of grid points in the gas-phase and in the washcoat and the aspect ratios are given in Table 5.7 for simple aspect ratio grid generation.

Table 5.7: Number of grid points and aspect ratios in the gas-phase and in the washcoat

	Number of grid points	Aspect ratio
Gas-phase	40	1.03
Washcoat	30	1.06

Adaptive gridding is performed from Case 1 to Case 5 for *η-approach* and RD-approach. It is exemplarily shown here in Case 1 for RD-approach. In Case 1, *RD-approach* simulations are initialized with 10 points in the gas-phase and 6 points in the washcoat. The simulation ended with 35 mesh points in the gas-phase and 41 mesh points in the washcoat as shown in Fig.5.20 and Fig.5.21.

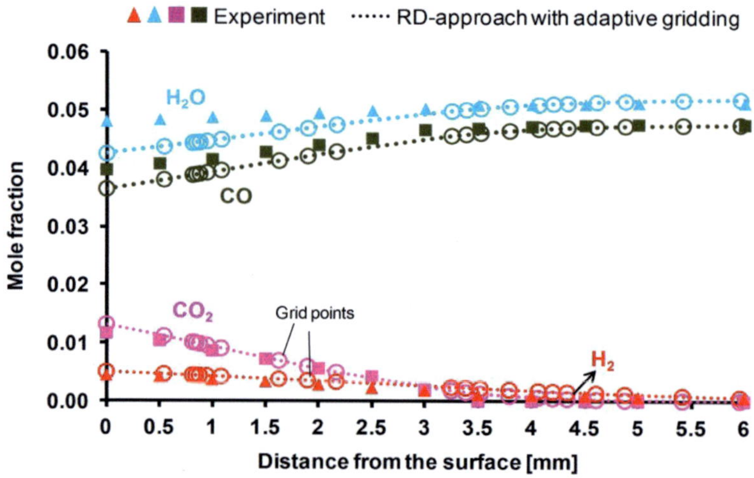

Figure 5.20: Species profiles in the gas-phase in WGS reaction at 873 K: the grids are generated with adaptive grid method

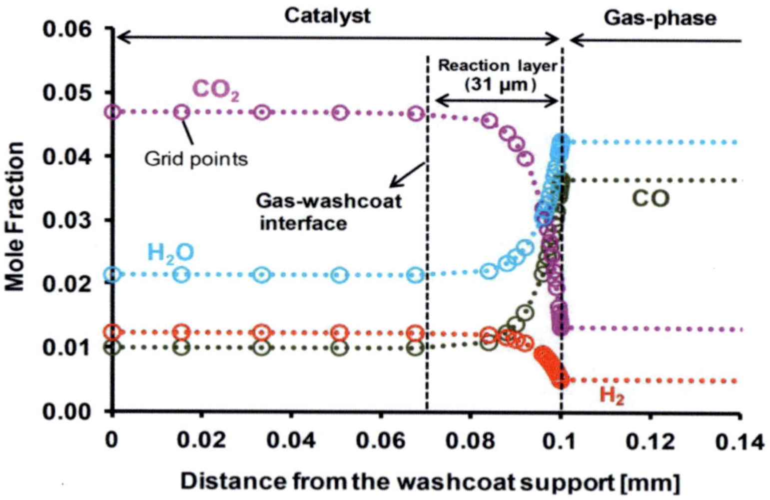

Figure 5.21: Species profiles in the washcoat in catalytic oxidation of CO at 873 K: the grids are generated with adaptive grid method

5.4 Conclusions

WGS and RWGS are investigated numerically in stagnation flow over a 100 µm catalytic disk. Simulations with the 1D SFR model predicts the experiments, well. According to the *η-approach*, *RD-approach* and *DGM* simulations, internal mass transfer limitations are very significant in the systems studied. Therefore, diffusion limitations inside the washcoat must be considered for accurate interpretation of the experimental data in case of thick catalyst layers.

The effect of external mass transfer limitations on conversion is rather small for the studied WGS and RWGS cases. Therefore, internal mass transfer limitations are the prominent reason for low activities in the studied WGS and RWGS cases.

The *DGM* simulations give identical species profiles with the *RD-approach* for washcoat applications due to low pressure gradients. Therefore, the species transport inside the washcoat due to pressure-driven convective flow is negligible as expected for a reaction with constant volume at first sight. However, the strongly different diffusion coefficients in the presence of hydrogen may have some effect on convective flow.

Decreasing the inlet flow velocity and increasing the reactor pressure results in an increase of conversion. Thinner washcoat layers along with the higher reactor pressures (3 atm) and lower inlet flow velocities (0.2-0.3 m/s) would result in a high CO consumption. In addition, if the mean pore diameter and porosity in the washcoat is increased, and tortuosity is decreased, internal mass transfer limitations are decreased significantly.

6 Partial Oxidation and Steam Reforming of Methane on Rh/Al_2O_3

In this chapter, catalytic partial oxidation (CPOX) and steam reforming (SR) of methane (CH_4) are numerically investigated in stagnation flow over a Rh/Al_2O_3 catalyst. Numerical simulations are applied based on the recent SFR experiments of Karakaya [98]. Possible reaction routes, internal mass and heat transport limitations and the effect of convective flow inside the catalyst are investigated. The effect of flow rates and pressure on internal and external mass transfer limitations and syngas production is investigated. In addition, boundary layer thicknesses and external mass transfer limitations in the gas-phase are discussed.

6.1 Theoretical background

Syngas, a mixture of carbon monoxide (CO) and hydrogen (H_2), is used as an important chemical intermediate in the chemical processes such as Fischer-Tropsch (FT) and methanol synthesis. In addition, it is an alternative fuel for the solid oxide fuel cells (SOFC), which can be used to generate power in small units or large scale power plants [123]. Today, syngas is mostly produced via steam reforming (SR) of methane (CH_4) (Eq. (6.1)), which is the main constituent of natural gas, in tubular reactors packed with supported Ni catalysts. This system has certain drawbacks such as low catalyst effectiveness factors, weak heat transport capabilities, large-scale operation and significant initial capital expenditures [112, 124]. Therefore, microchannel reactors with noble metal catalysts have been investigated recently as an alternative to tubular reactors for SR of CH_4 [125-130]. Microchannel reactors with rhodium catalysts offer enhanced heat and mass transfer, safe control in explosive regime, high surface area, low pressure drop and short residence time (10ms or less) [111, 113, 131]. In addition, the process is 100-1000 fold smaller than bulky reformers.

$$CH_4 + H_2O \rightarrow CO + 3H_2 \qquad \Delta H_r = +205.9\ kJ/mol \tag{6.1}$$

Catalytic partial oxidation (CPOX) of CH_4 (Eq. (6.2)) is an attractive alternative fuel processing method to the large SR reactors [131]. The process is well suited for small scale systems, such as foam catalysts, monolithic reactors and micro-reformers. It is simple and no humidification of the inlet stream is necessary [64]. The H_2/CO ratio of syngas from CPOX of CH_4 is also more suitable as feedstock for methanol synthesis and the Fischer-Tropsch reaction, compared to SR processes [120]. Recently, there is also an interest on using a two-stage process for obtaining a stabilized catalytic combustion at power generation applications [132]. In the first catalytic fuel-rich step, partial oxidation of CH_4 occurs where CO and H_2 are formed. In the second phase, the formed H_2 stabilizes the lean homogenous combustion. Rhodium is an active and stable noble metal for CPOX applications as well.

$$CH_4 + 1/2\,O_2 \rightarrow CO + 2H_2 \qquad \Delta H_r = -35.9\ kJ/mol \tag{6.2}$$

Understanding the chemical and physical steps in CPOX and SR of CH_4 for catalytic reactors will help to explore the reactor conditions and optimize the catalyst [131]. For this purpose, the reaction mechanism and product development in the catalyst have been investigated in the literature for the last two decades. Direct and indirect reaction mechanisms are proposed for the partial oxidation of CH_4 on Rh catalysts. Some studies suggested a single step process (direct mechanism), which assumes that syngas is primarily formed by partial oxidation [133-136]. On the other hand, other studies have postulated a two-step mechanism, where CH_4

reacts initially with O_2 to form CO_2 and H_2O (total oxidation) followed by steam and dry reforming of CH_4 [132, 137-141]. Recently, microkinetic studies have also been employed for SR of CH_4 on Rh catalysts [98, 142-144] . Maestri et al. [143] proposed that CH_4 and H_2O convert to CO and H_2, and then CO reacts with H_2O leading to CO_2 and H_2. Since inclusion of a porous layer on the surface of the catalytic reactors in CPOX and SR of CH_4 is a common application, the impact of internal mass and heat transport limitations on the system should also be investigated [64, 145, 146]. The effect of pressure and flow rates on the internal and external mass transport limitations and syngas production should also be analyzed for a complete understanding.

6.2 Surface Reaction Mechanism for Catalytic Partial Oxidation and Steam Reforming of Methane over Rh/Al_2O_3

The surface reaction mechanism used in this chapter is taken from Karakaya [98] where 48 irreversible surface reactions with 7 gas-phase and 13 surface species are written to describe the partial oxidation as well as steam reforming of CH_4. The surface reaction mechanism is given in the appendix in Table B.1. The reaction mechanism was developed based on the experimental data that confirms possible indirect reaction paths. CO_2 and H_2O are formed via direct oxidation of methane. SR, WGS, RWGS and methanation reactions are also considered to describe the indirect path of H_2 and CO formation.

Based on the kinetics scheme, main reaction path of SR differ depending on the temperature. At low temperature (773 K) reaction is sensitive to CO, H_2O species and their reactions where WGS reaction path is dominant. Adsorption, desorption and dehydrogenation steps of CH_4 are dominant for production of CO. At high temperature regimes, formation of CO is sensitive to adsorption and desorption steps of CH_4. H_2O concentration does not have a big influence on the reaction rate and is independent of H_2O concentration. The rate determining step is related to the methane pyrolysis reaction step.

6.3 Results and Discussion

6.3.1 Cases Studied

In this chapter, the experimental stagnation-flow reactor data of [98] is used to investigate CPOX and SR of CH_4 over Rh/Al_2O_3. In this respect, a slightly lean/stoichiometric condition (C/O=1.03) for the partial oxidation of CH_4 is studied at 873K, in Case 1. In Case 2, the stoichiometric condition for the partial oxidation of CH_4 is examined at 973K. Case 3 and Case 4 consider the stoichiometric and fuel-rich conditions for the total oxidation of CH_4 at 973 K, respectively. In addition, a slightly rich condition (close to the stoichiometry) for the partial oxidation of CH_4 is considered at 1023 K, in Case 5. The conditions from Case 1 to Case 5 are summarized in Table 6.1. Subsequently, SR of CH_4 is investigated at 973 K and 1023 K (Case 6 and Case 7). The conditions for the SR of CH_4 are given in Table 6.2.

Table 6.1: Reaction conditions for CPOX of CH_4

	T_{disc} (K)	T_{inlet} (K)	CH_4 (% vol.)	O_2 (% vol.)	C/O -	Ar (% vol.)	Inlet vel. (cm/s)	Reactor pres. (mbar)
Case 1	873	313	5.30	2.57	1.03	87.82	51	500
Case 2	973	313	5.32	2.78	0.99	91.90	51	500
Case 3	973	313	5.20	4.90	0.53	89.90	51	500
Case 4	973	313	4.38	7.80	0.28	87.82	51	500
Case 5	1023	313	5.20	2.83	0.93	91.51	51	500

Table 6.2: Reaction conditions for SR of CH_4

	T_{disc} (K)	T_{inlet} (K)	CH_4 (% vol.)	H_2O (% vol.)	S/C -	Ar (% vol.)	Inlet vel. (cm/s)	Reactor pres. (mbar)
Case 6	973	423	5.06	5.38	1.06	89.56	71	500
Case 7	1008	423	5.16	5.38	1.04	89.46	71	500

6.3.2 Input data for numerical simulations

The inlet conditions for the numerical simulations are based on the experimental conditions. The inlet velocity is taken as 51 cm/s and 71 cm/s for CPOX and SR of CH_4, respectively. The finite gap between the inlet and catalytic surface is 3.9 cm. The reactor inlet temperature is taken as 313 K and 423 K for CPOX and SR cases, respectively.

In this chapter, the simulations are performed with three different transport models, i.e., with the *η-approach*, *RD-approach* and *DGM*. CH_4 is chosen as the rate-limiting species for the *η-approach* simulations. *η-approach* results are examined by choosing O_2 as the rate-limiting species for Cases 1-5, as well. The thickness, mean pore diameter, tortuosity and porosity of the washcoat are the parameters that are used in the *η-approach*, *RD-approach* and *DGM* simulations. In *DGM*, particle diameter is also needed. The values for these parameters are given in Table 6.3.

Table 6.3: The parameters used in the surface models

Reaction Case	Thickness of the washcoat (μm)	Mean pore diameter (nm)	Porosity (%)	Tortuosity	Particle diameter (nm) (*DGM* only)
Case 1	100	10	40	8	100
Case 2	100	10	60	8	100
Case 3	100	10	40	8	100
Case 4	100	10	40	8	100
Case 5	100	10	40	8	100
Case 6	100	10	40	8	100
Case 7	100	10	40	8	100

Simple aspect ratio is used for grid generation. In addition, only indirect coupling of the washcoat and gas-phase is applied in the simulations.

6.3.3 Catalytic Partial Oxidation of Methane

The experimental and simulation results for Case 1 (C/O=1.03) are given in Fig. 6.1. According to the experiments, synthesis gas yield is low at this condition. Total oxidation products (CO_2 and H_2O) are the main products at the catalyst surface. The species boundary layer in the gas-phase is around 6 mm (Fig. 6.1), relative to the external catalyst surface. In this case, η-approach, *RD-approach* and *DGM* simulations show relatively good agreement with the experiments. η-approach predicts reactant's and total oxidation product's mole fractions closely to the experiments, when O_2 is chosen as the rate-limiting species. However, the model does not predict any syngas production in this case (Appendix C, Fig. C1). *RD-approach* and *DGM* simulations give an insight to realize the physical and chemical processes (reaction routes) in the washcoat (Fig. 6.2). According to the DGM simulation, the pressure difference between the gas-washcoat interface and the washcoat support side is low for Case 1 (Table 6.4). Therefore, *DGM* yields identical species profiles with the *RD-approach* (Fig. 6.2).

Table 6.4: The pressure difference in the washcoat in CPOX cases

	Case 1	Case 2	Case 3	Case 4	Case 5
Pressure difference (Pa)	494	440	104	45	403

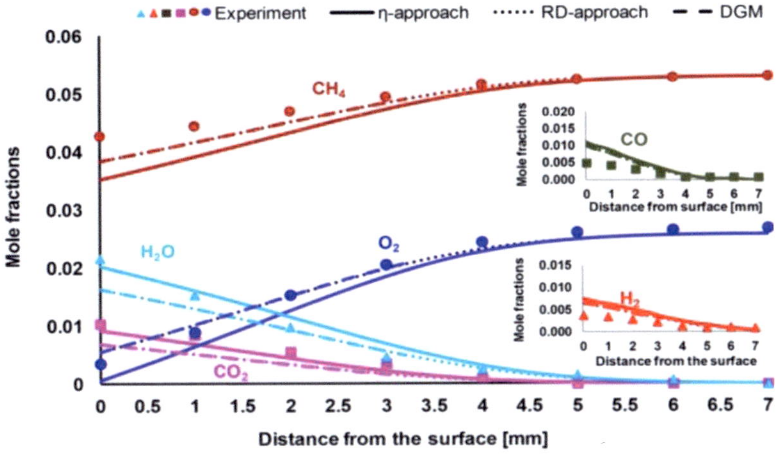

Figure 6.1: Experimental and simulation results for the species profiles in CPOX of CH4 at 873 K, indirect coupling of the gas-phase and washcoat is used, grids are generated with simple aspect ratio (Case 1, C/O=1.03, stoichiometric for partial oxidation)

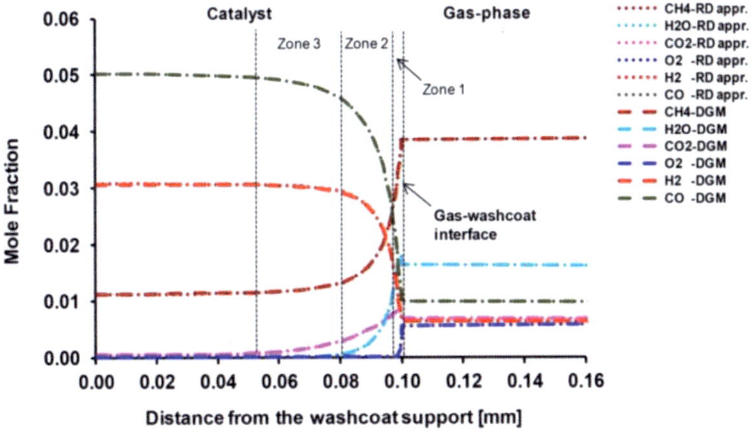

Figure 6.2: Species mole fractions inside the porous washcoat layer in CPOX of CH4 at 873 K, indirect coupling of the gas-phase and washcoat is used, grids are generated with simple aspect ratio (Case 1, C/O=1.03, stoichiometric for partial oxidation)

In Fig. 6.3, only the first and second reaction zones in the washcoat from Fig. 6.2 are considered for detailed explanation. In Zone 1, there is a thin total oxidation zone near the external catalyst surface. After this thin total oxidation zone, mainly SR of CH_4 occurs in Zone 2. Dry reforming (DR) of CH_4 (Eq. (6.3)) is observed simultaneously in this zone as well, but to a much lesser extent. In Zone 3 (Fig. 6.2), only a slight DR of CH_4 is observed. After Zone 3, there is not any reaction in the rest of the washcoat. The species composition at the 0.05 mm of the washcoat is used in DETCHEM^{EQUIL} code to realize if the composition has reached the thermodynamic equilibrium. Further, DETCHEM^{EQUIL} code calculations show that the species composition has reached the equilibrium at 0.05 mm of the washcoat (Table 6.5).

$$CH_4 + CO_2 \rightarrow 2CO + 2H_2 \qquad \Delta H_r = +247.0\ kJ/mol \tag{6.3}$$

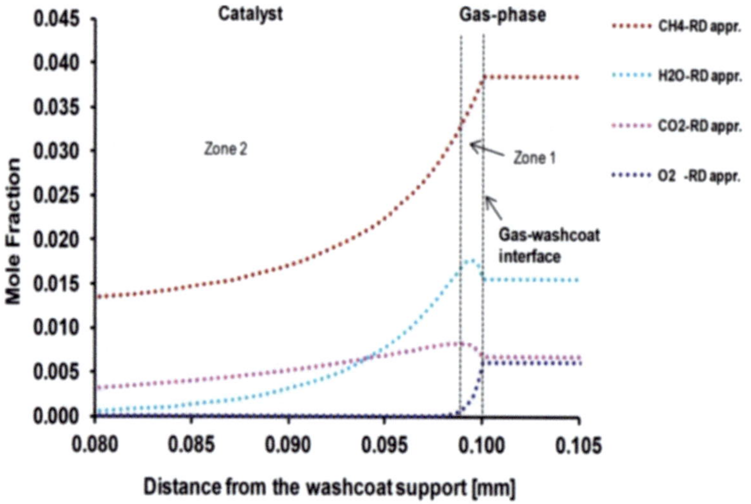

Figure 6.3: Reaction routes in the first and second zone of the washcoat (Case 1, C/O=1.03, stoichiometric for partial oxidation)

Table 6.5: Equilibrium composition between 0-0.05 mm in the washcoat (Case 1, C/O=1.03, stoichiometric for partial oxidation)

Species	CH_4	O_2	H_2O	CO_2	H_2	CO	AR
Mole Fraction	1.11 E-02	6.70 E-21	6.20 E-05	2.70 E-04	3.09 E-02	5.02 E-02	9.07 E-01

Fig. 6.4 shows the experimental and simulation results for Case 2 (C/O=0.99). In Case 2, species boundary layer in the gas-phase is again around 7 mm. Experiments show that O_2 is almost completely consumed on the surface. The main products are synthesis gas and total oxidation products (CO_2 and H_2O). η-approach, *RD-approach* and *DGM* simulations show relatively good agreement with the experiments. There is a slight deviation for the H_2O production and O_2 consumption predictions. However, these deviations might also be due to slight inaccuracies in the experiments. η-approach cannot predict syngas production accurately, when O_2 is chosen as the rate-limiting species (Appendix C, Fig. C2).

In Fig. 6.5, only 30 μm of the washcoat is shown for Case 2, because the reactions occur only in this section. According to *RD-approach and DGM* simulations, total oxidation is a weak process due to too little amount of O_2 inside the catalyst (Fig. 6.5) at steady state. There exist SR and DR of CH_4 inside the first reaction zone of the washcoat. However, DR occurs in a much lesser extent than SR. There is just a slight DR process within the second reaction zone. Internal mass transfer limitations are observed for this case as well. The whole reaction layer is around 30 μm. The rate-limiting process is the internal diffusion. *η-approach* yields Φ = 27.4 and η = 0.04, respectively, confirming the diffusion limitations. Pressure difference in the washcoat is 440 Pa (Table 6.4).

DETCHEM^{EQUIL} code shows that the chemical composition already reaches thermodynamic equilibrium at the 70 μm of the washcoat as given in Table 6.6.

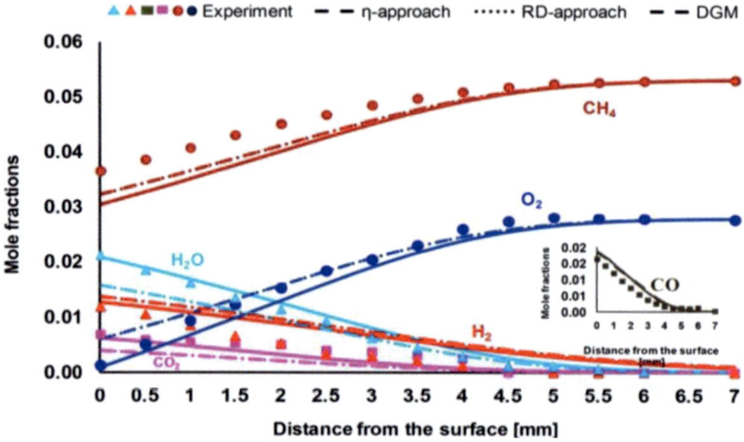

Figure 6.4: Experimental and simulation results for the species profiles in CPOX of CH4 at 973 K, indirect coupling of the gas-phase and washcoat is used, grids are generated with simple aspect ratio (Case 2, C/O=0.99, stoichiometric for partial oxidation)

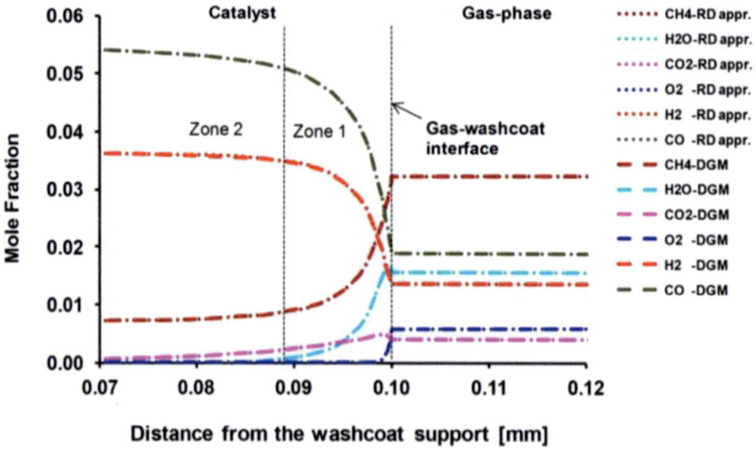

Figure 6.5: Species mole fractions inside the porous washcoat layer in CPOX of CH4 at 973 K, indirect coupling of the gas-phase and washcoat is used, grids are generated with simple aspect ratio (Case 2, C/O=0.99, stoichiometric for partial oxidation)

Table 6.6: Equilibrium composition in the washcoat at 70 µm (Case 2, C/O=0.99, stoichiometric for partial oxidation)

Species	CH_4	O_2	H_2O	CO_2	H_2	CO	AR
Mole Fraction	6.60 E-03	1.61 E-20	8.38 E-06	2.01 E-05	3.73 E-02	5.50 E-02	9.01 E-01

In Case 3, CH_4 consumption rate is increased, compared to Case 2, due to increased amount of oxygen (Fig. 6.6). Therefore, more total oxidation products are obtained. The amount of synthesis gas products is decreased. *η-approach* predicts the experiments reasonably well. *RD-approach* and *DGM* simulations predict the experiments well. *η*-approach cannot predict CO production accurately, when O_2 is chosen as the rate-limiting species (Appendix C, Fig. C3). Pressure difference in the washcoat is low in this case as well (Table 6.4). According to the *RD-approach* and *DGM* simulations, there are complex processes inside the washcoat (Fig. 6.7). The reaction layer is divided into 3 zones in Fig. 6.7. The first zone, which is adjacent to the external catalyst surface, shows a thin reaction layer where total oxidation occurs. In the second zone, there is SR of CH_4, where CH_4 and H_2O are consumed, CO and H_2 are produced. CO_2 is still formed in the second zone due to WGS reaction. In the third zone, CO_2 is not formed anymore. The little amount of remaining CH_4 reacts with H_2O (SR) to yield synthesis gas. Surface reactions are fast at this condition as well, and the rate-limiting process is again the internal diffusion. *η-approach* yields Φ = 30.5 and η = 0.03, respectively. DETCHEM^{EQUIL} code shows that the chemical composition reaches thermodynamic equilibrium at the 80 µm of the washcoat as given in Table 6.7.

Table 6.7: Equilibrium composition in the washcoat at 80 µm (Case 3, C/O=0.53, stoichiometric for total oxidation)

Species	CH_4	O_2	H_2O	CO_2	H_2	CO	AR
Mole Fraction	2.10 E-06	1.13 E-20	1.22 E-02	2.01 E-05	3.18 E-02	4.12 E-02	8.89 E-01

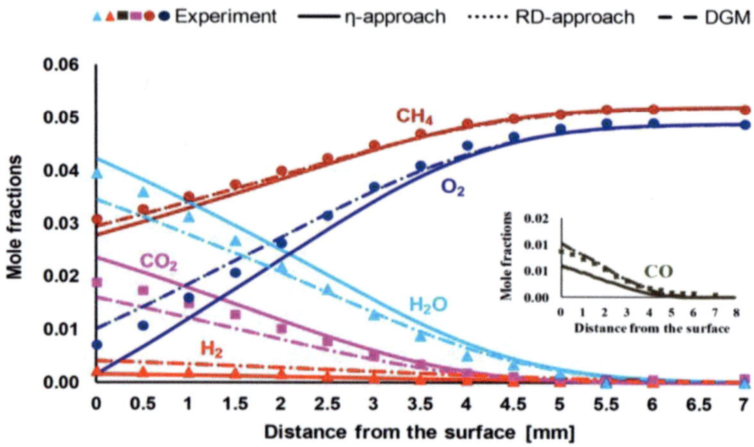

Figure 6.6: Experimental and simulation results for the species profiles in CPOX of CH4 at 973 K, indirect coupling of the gas-phase and washcoat is used, grids are generated with simple aspect ratio (Case 3, C/O=0.53, stoichiometric for total oxidation)

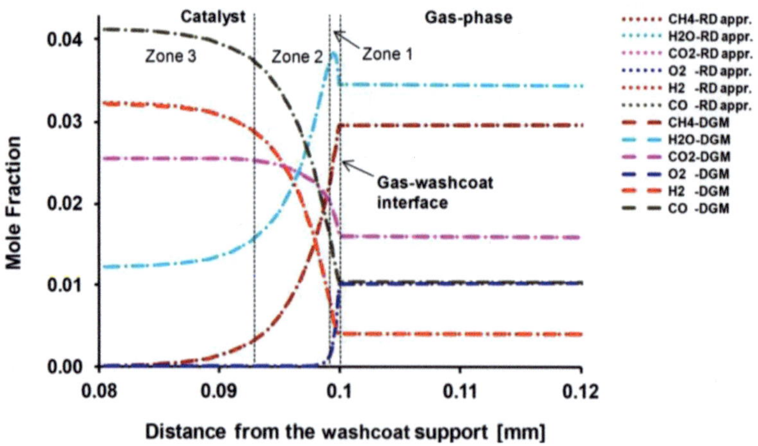

Figure 6.7: Species mole fractions inside the porous washcoat layer in CPOX of CH4 at 973 K, indirect coupling of the gas-phase and washcoat is used, grids are generated with simple aspect ratio (Case 3, C/O=0.53, stoichiometric for total oxidation)

The experimental and simulation results for Case 4 are given in Fig. 6.8. According to the experiments, the main products on the surface are the total oxidation products (CO_2 and H_2O). Syngas products are not obtained on the surface at this fuel-rich (for total oxidation) case. The species boundary layer in the gas-phase is around 7 mm (Fig. 6.8). In this case, *η-approach*, *RD-approach* and *DGM* simulations predict the experiments well. η-approach predicts the experiments reasonably, when O_2 is chosen as the rate limiting species (Appendix C, Fig. C4). The low pressure difference inside the washcoat (Table 6.4) is again the reason for identical species profiles from the *RD-approach* and *DGM* simulations. *RD-approach* and *DGM* simulations reveal that there is a total oxidation zone in the washcoat (zone 1 in Fig. 6.9), near the external catalyst surface. After this total oxidation zone, there is the SR of CH_4 (zone 2 in Fig. 6.9). Since there is not any oxygen left and CO_2 is still formed, WGS occurs as well. SR and WGS occur simultaneously in the entire Zone 2. The total reaction layer (Zone 1 and Zone 2 together) is around 15 μm. Surface reactions are fast and internal mass transfer limitations are observed. *η-approach* yields Φ = 32.2 and η = 0.03, respectively. The species composition reaches thermodynamic equilibrium at the 85 μm of the washcoat as given in Table 6.8.

Case 5 considers the CPOX of CH_4 for C/O=0.93 at 1023 K. CH_4 is converted more in Case 5 compared to Case 2 due to increased surface temperature (Fig. 6.10). There is a slight increase of the synthesis gas products compared to Case 2. η-approach, RD-*approach* and *DGM* simulations show good agreement with the experiments again. η-approach cannot predict the experiments, when O_2 is chosen as the rate limiting species (Appendix C, Fig. C5). According to the *RD-approach* and *DGM* simulations total oxidation inside the washcoat is weak, due to little amount of O_2 inside the catalyst (Fig.6.11). Total reaction layer inside the washcoat decreases compared to Case 3 due to increased surface temperature (increased surface reaction rates). SR and DR processes simultaneously take place within the first zone of the washcoat. However, SR is already the dominant process. DR occurs slightly within the second zone. However, it is a weak process. In this case, *η-approach* yields Φ = 32.5 and η = 0.03, respectively. The species composition reaches thermodynamic equilibrium at the 70 μm of the washcoat as given in Table 6.9.

Table 6.8: Equilibrium composition in the washcoat at 85 μm (Case 4, C/O=0.28, fuel-rich for total oxidation)

Species	CH_4	O_2	H_2O	CO_2	H_2	CO	AR
Mole Fraction	2.19 E-09	4.61 E-20	5.32 E-02	5.17 E-02	1.00 E-02	6.00 E-03	8.79 E-01

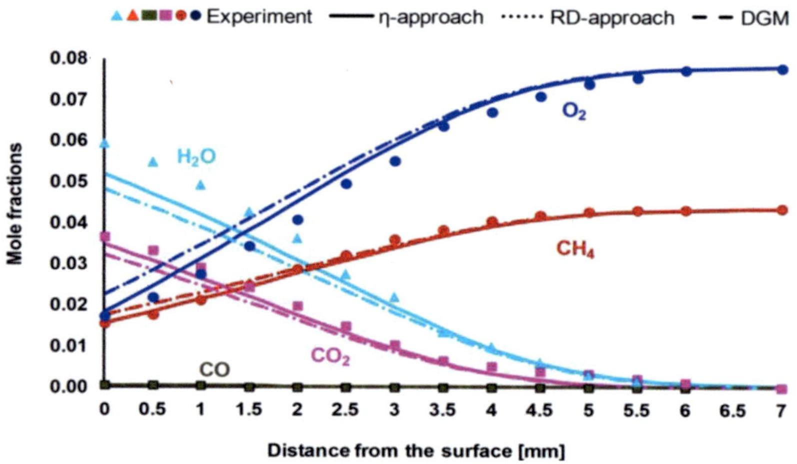

Figure 6.8: Experimental and simulation results for the species profiles in CPOX of CH_4 at 973 K, indirect coupling of the gas-phase and washcoat is used, grids are generated with simple aspect ratio (Case 4, C/O=0.28, fuel-rich for total oxidation)

Table 6.9: Equilibrium composition in the washcoat at 70 μm (Case 5, C/O=0.93, slightly lean for partial oxidation)

Species	CH_4	O_2	H_2O	CO_2	H_2	CO	AR
Mole Fraction	6.00 E-03	1.65 E-20	2.45 E-06	4.58 E-06	3.80 E-02	5.40 E-02	9.02 E-01

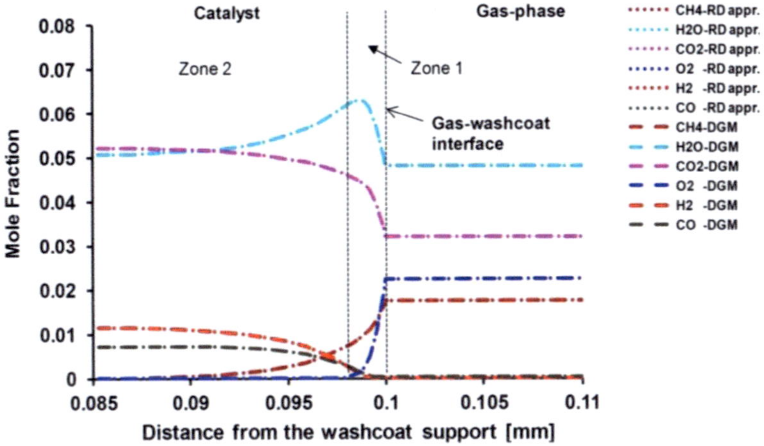

Figure 6.9: Species mole fractions inside the porous washcoat layer in CPOX of CH4 at 973 K, indirect coupling of the gas-phase and washcoat is used, grids are generated with simple aspect ratio (Case 4, C/O=0.28, fuel-rich for total oxidation)

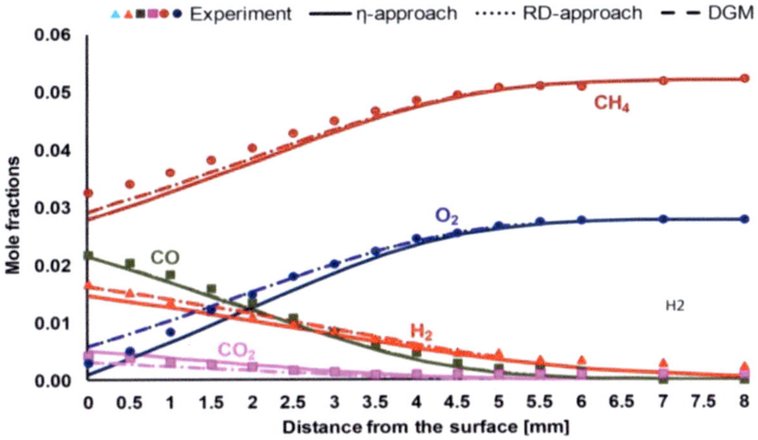

Figure 6.10: Experimental and simulation results for the species profiles in CPOX of CH4 at 1023 K, indirect coupling of the gas-phase and washcoat is used, grids are generated with simple aspect ratio (Case 5, C/O=0.93, slightly lean for partial oxidation)

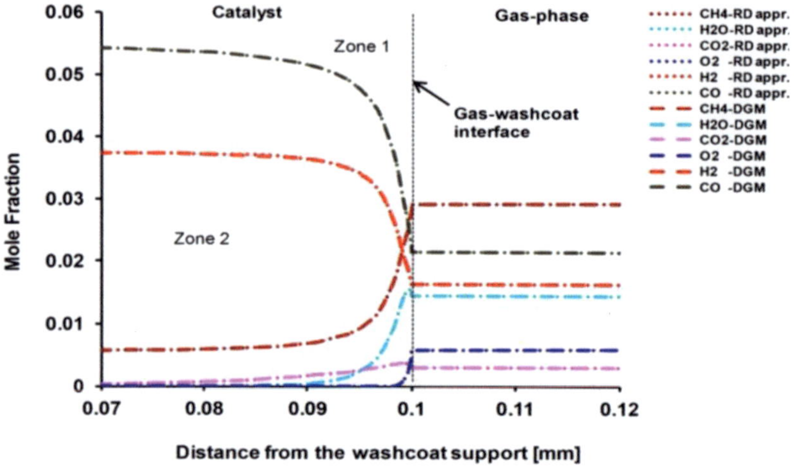

Figure 6.11: Species mole fractions inside the porous washcoat layer in CPOX of CH4 at 1023 K, indirect coupling of the gas-phase and washcoat is used, grids are generated with simple aspect ratio (Case 5, C/O=0.93, slightly lean for partial oxidation)

6.3.4 Steam Reforming of Methane

Case 6 considers the SR of CH_4 at 973K. The species boundary layer in the gas-phase is around 9 mm (Fig.6.12). It is observable that the CO/H_2 ratio on the surface obtained from SR of CH_4 at 973 K differs from the CO/H_2 ratio obtained from CPOX of CH_4 at 973K (Fig.6.12). η-approach, *RD-approach* and *DGM* predict the experiments well. According to the DGM simulation, the pressure difference between the gas-washcoat interface and the washcoat support side is low for Case 6 (Table 6.10). Therefore, *DGM* yields identical species profiles with the *RD-approach*. The reaction layer inside the washcoat is divided into two zones (Fig.6.13). There is a very slight WGS kinetics within the first reaction zone. However, the driving process here is SR of CH_4, where most of the CH_4 and H_2O are converted to synthesis gas. In the second zone, there is no more WGS kinetics, but a slight SR of CH_4. Internal mass transfer resistances are observable in the washcoat due to fast surface reactions. Therefore, the whole reaction layer is only 20 µm. *η-approach* yields Φ = 26.46 and η = 0.0378, respectively, confirming the strong diffusion limitation.

Table 6.10: The pressure difference in the washcoat in SR cases

	Case 6	Case 7
Pressure difference (Pa)	472	464

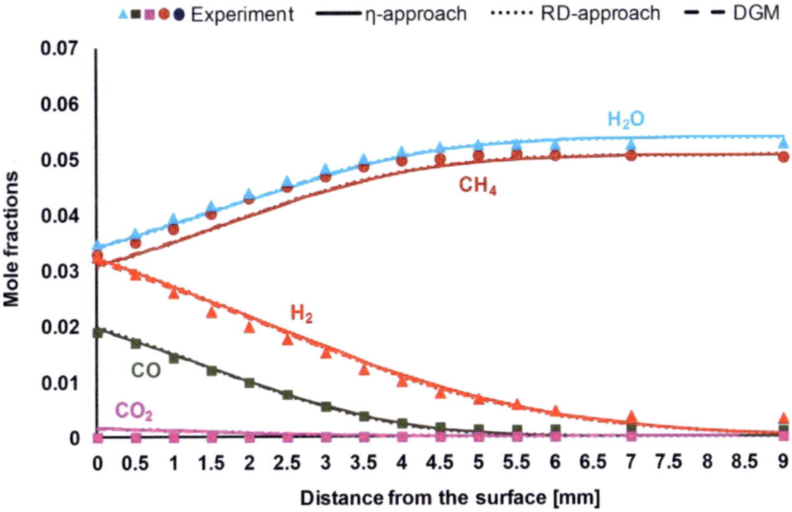

Figure 6.12: Experimental and simulation results for the species profiles in SR of CH_4 at 973 K, indirect coupling of the gas-phase and washcoat is used, grids are generated with simple aspect ratio (Case 6)

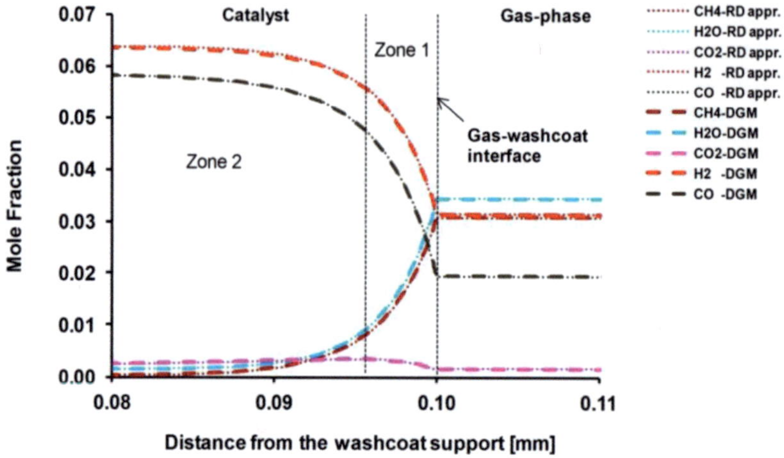

Figure 6.13: Species mole fractions inside the porous washcoat layer in SR of CH4 at 973 K, indirect coupling of the gas-phase and washcoat is used, grids are generated with simple aspect ratio (Case 6)

The chemical composition reaches thermodynamic equilibrium at the 80 µm of the washcoat according to the DETCHEM^{EQUIL} code calculations (Table 6.11).

Table 6.11: Equilibrium composition in the washcoat at 80 µm (Case 6, steam reforming at 973 K)

Species	CH_4	O_2	H_2O	CO_2	H_2	CO	AR
Mole Fraction	1.81 E-04	3.97 E-21	1.56 E-03	2.22 E-03	6.39 E-02	5.59 E-02	8.76 E-01

Finally, Case 7 considers SR of CH_4 at 1008K. An increased reaction rate is observed for CH_4 and O_2, compared to Case 6, due to increased surface temperature (Fig.6.14). Therefore, a higher synthesis gas yield is obtained. *η-approach, RD-approach* and *DGM* predict the experiments well.

The pressure difference in the washcoat is low in this case as well (Table 6.10). Fig.6.15 shows that the reaction layer is just within the first 14-15 µm relative to the external catalyst surface. There is only SR of methane within the whole reaction layer. *η-approach* yields Φ = 30.1 and η = 0.03, respectively.

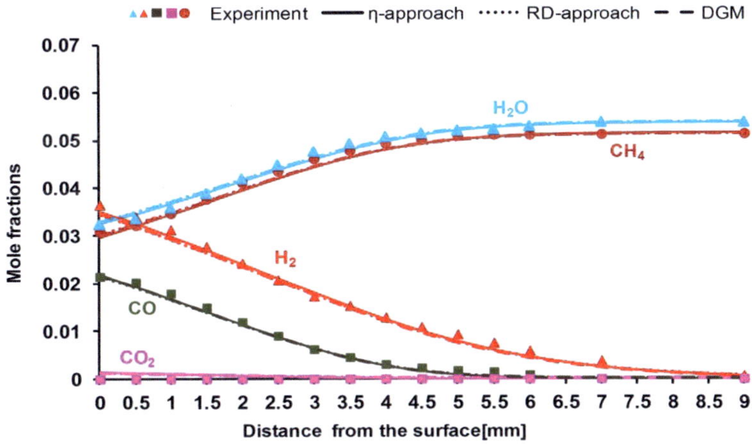

Figure 6.14: Experimental and simulation results for the species profiles in SR of CH4 at 1008 K, indirect coupling of the gas-phase and washcoat is used, grids are generated with simple aspect ratio (Case 7)

The chemical composition reaches thermodynamic equilibrium at the 85 µm of the washcoat according to the DETCHEM^{EQUIL} code calculations (Table 6.12).

Table 6.12: Equilibrium composition in the washcoat at 80 µm (Case 7, steam reforming at 1008 K)

Species	CH_4	O_2	H_2O	CO_2	H_2	CO	AR
Mole Fraction	1.08 E-04	5.21 E-21	1.16 E-03	1.49 E-03	6.58 E-02	6.02 E-02	8.76 E-01

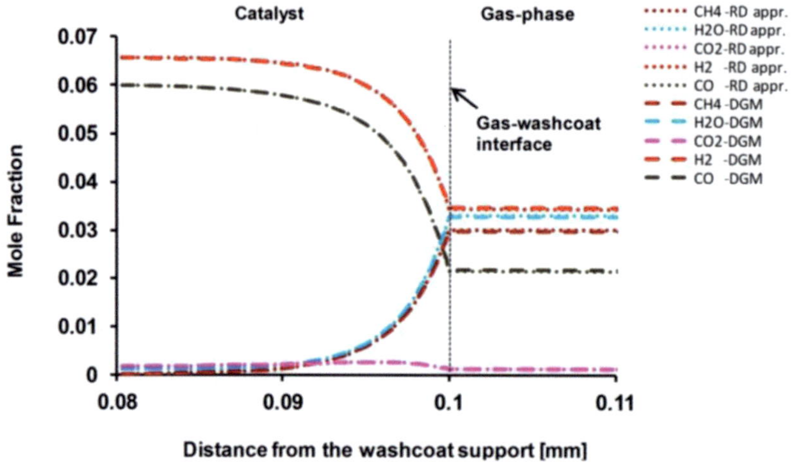

Figure 6.15: Species mole fractions inside the porous washcoat layer in SR of CH4 at 1008 K, indirect coupling of the gas-phase and washcoat is used, grids are generated with simple aspect ratio (Case 7)

6.3.5 The Effect of Pressure and Flow Rates on External and Internal Mass Transfer Limitations and Syngas Production in CPOX and SR of CH_4

In this section, the effect of the pressure and flow rates on syngas production is investigated for CPOX and SR of CH_4. The inlet mole fractions, the surface and inlet temperatures for CPOX and SR of CH_4 are taken from Case 2 and Case 6, respectively. The simulations are initially performed with varying pressures from 0.5 to 3 bar, and varying inlet velocity from 0.2 to 0.9 m/s. External mass transfer limitations are discussed based on Damkohler (*Da*) number. Internal mass transfer limitations are discussed based on effectiveness factor (η).

CPOX of CH_4: Fig.6.16 shows that external mass transfer limitations become important with the increasing reactor pressure and decreasing inlet flow velocity. *Da* number reaches 8.1 at 3 atm pressure and 0.2 m/s inlet velocity. It reaches 3.8 at 0.5 atm pressure and 0.9 m/s.

Fig.6.17 shows that internal mass transfer limitations decrease significantly with the increasing reactor pressure and decreasing inlet flow velocity. η reaches 0.051 at 3 atm pressure and 0.2 m/s inlet velocity. It reaches 0.036 at 0.5 atm pressure and 0.9 m/s.

The mole fraction of H_2 at the surface increases with the increasing reactor pressure and decreasing inlet flow velocity (Fig.6.18). Mole fraction of H_2 at the surface is 0.032 at 3 atm pressure and 0.2 m/s inlet velocity. It is 0.007 at 0.5 atm pressure and 0.9 m/s inlet flow velocity.

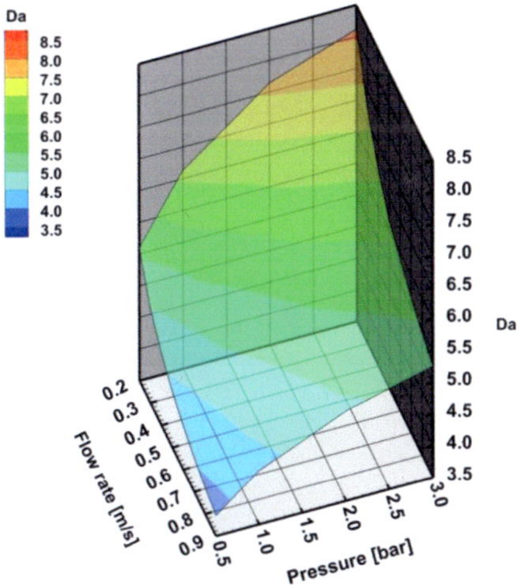

Figure 6.16: Da number with varying reactor pressure and varying inlet flow velocity in CPOX of CH4 at 973K (C/O=0.99), the results are obtained with effectiveness factor approach

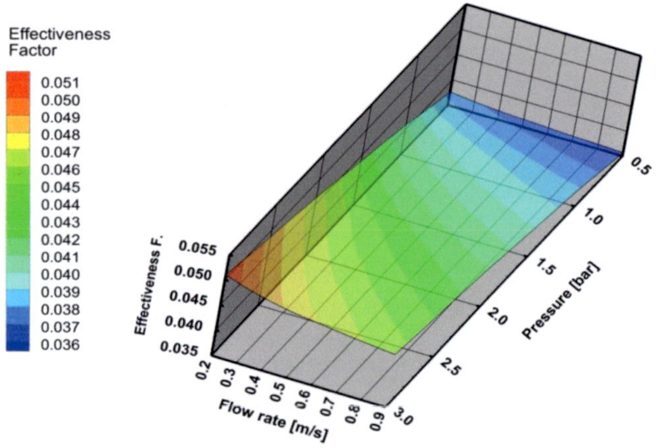

Figure 6.17: η number with varying reactor pressure and varying inlet flow velocity in CPOX of CH_4 at 973 K (C/O=0.99), the results are obtained with effectiveness factor approach

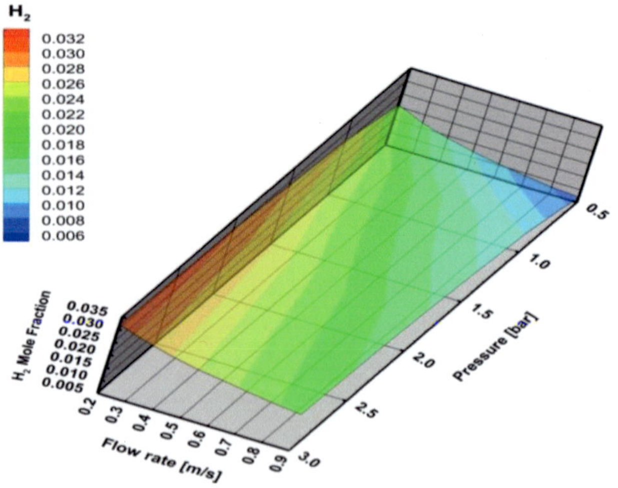

Figure 6.18: H_2 mole fraction at the surface with varying reactor pressure and varying inlet flow velocity in CPOX of CH4 at 973 K(C/O=0.99), the results are obtained with effectiveness factor approach

The results that are given above indicate that syngas yield in CPOX of CH_4 increases with the increasing reactor pressure and decreasing inlet flow velocity.

SR of CH_4: Fig.6.19 shows that external mass transfer limitations become important with the increasing reactor pressure and decreasing inlet flow velocity in SR of CH_4. *Da* number is obtained 5.91 at 3 atm pressure and 0.2 m/s inlet velocity. It is obtained 2.65 at 0.5 atm pressure and 0.9 m/s.

Fig.6.20 shows that internal mass transfer limitations decrease significantly with the increasing reactor pressure and decreasing inlet flow velocity. η reaches 0.0549 at 3 atm pressure and 0.2 m/s inlet velocity. It reaches 0.0375 at 0.5 atm pressure and 0.9 m/s.

The mole fraction of H_2 at the surface increases with the increasing reactor pressure and decreasing inlet flow velocity (Fig.6.21). Mole fraction of H_2 at the surface is 0.0545 at 3 atm pressure and 0.2 m/s inlet velocity. It is 0.0295 at 0.5 atm pressure and 0.9 m/s inlet flow velocity. These results indicate that syngas production can be increased with higher reactor pressure and lower inlet flow velocity.

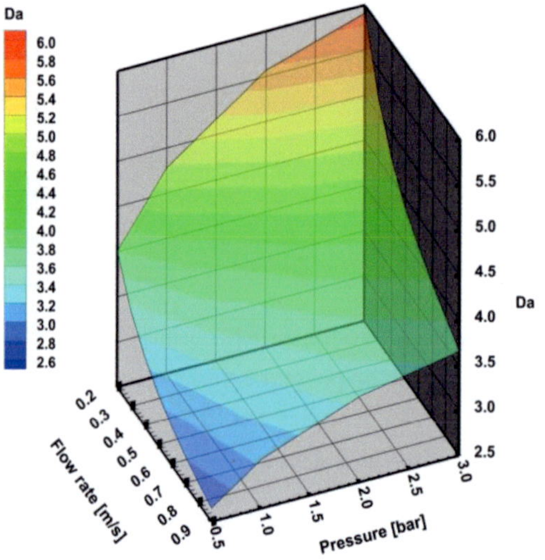

Figure 6.19: Da number with varying reactor pressure and varying inlet flow velocity in SR of CH_4, the results are obtained with effectiveness factor approach

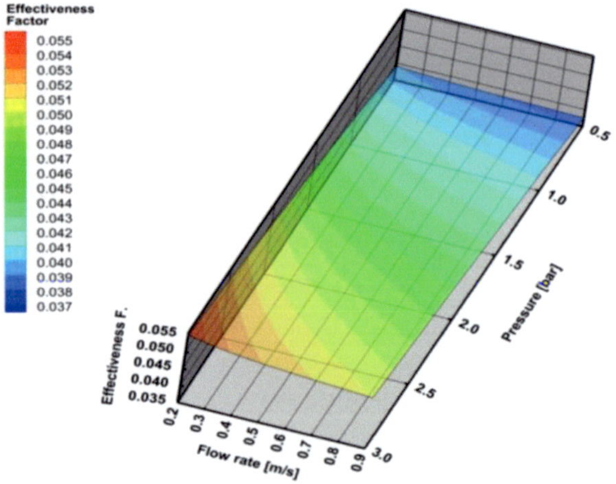

Figure 6.20: η number with varying reactor pressure and varying inlet flow velocity in SR of CH_4 at 973 K, the results are obtained with effectiveness factor approach

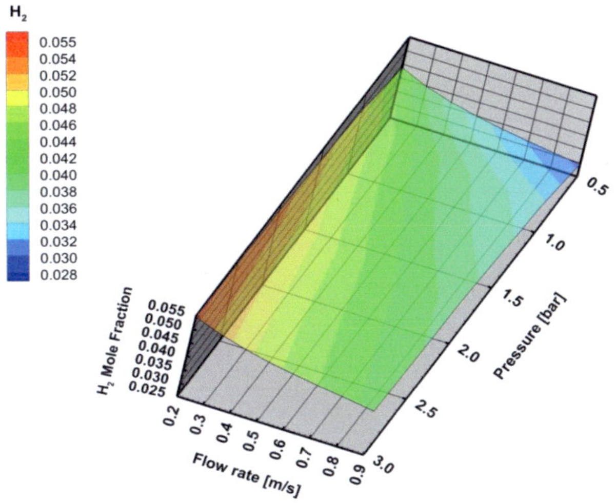

Figure 6.21: H_2 mole fraction at the surface with varying reactor pressure and varying inlet flow velocity in CPOX of CH_4 at 973 K

6.3.6 The Effect of Heat Transport Limitations in the Washcoat

In order to study the effect of the heat transport limitations in the washcoat, the experimental configuration of SFR which was used by Karakaya [98] should be examined. In the experiments of [98], the resistive heater (FeCrAl alloy) is used for supplying the required heat to the washcoat. There is the ceramic support between the resistive heater and the washcoat (Fig.6.22).

Heat transport between the resistive heater and the gas/washcoat interface involves different contributions. The following energy conservation equations are coupled in this study only with the *RD-approach*.

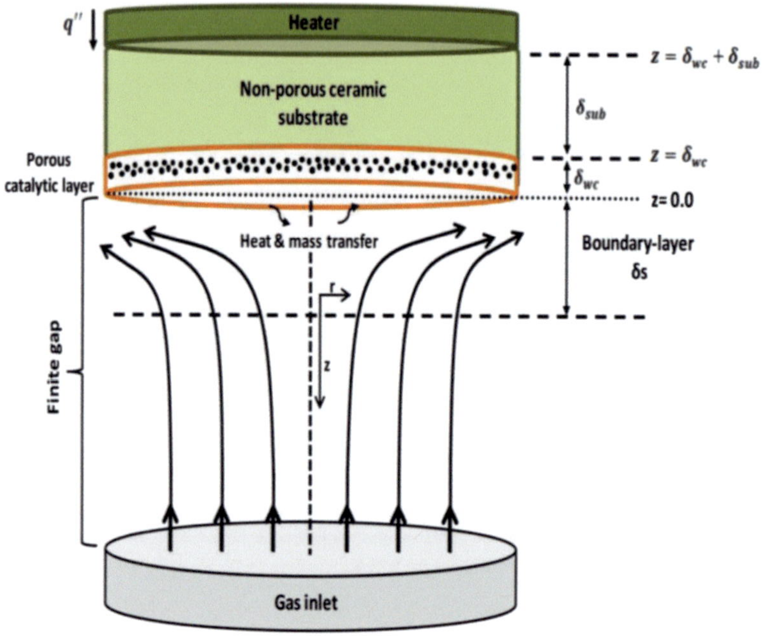

Figure 6.22: Fig.6.22. Experimental configuration of SFR which was used by [98]

Heat flux from the heater to the ceramic support is calculated from the resistive heating. Energy equation for the ceramic substrate is given as

$$\rho_{cr} C_{p,cr} \frac{\partial T_{cr}}{\partial t} = \lambda_{cr} \frac{\partial^2 T_{cr}}{\partial z^2} \tag{6.4}$$

where the left hand side represents the energy storage in the ceramic substrate. The right hand side represents the conduction of energy along the substrate. Energy equation inside the washcoat is given from Eq. (2.104).

In *RD-approach*, it is assumed that the diffusive mass flux in the washcoat is due to concentration gradient [20]. Here, the approach is extended by assuming that the diffusive mass flux in the washcoat is due to both concentration and temperature gradient. Therefore, diffusive mass flux is given as

$$j_i^w = -\left(D_{i,eff}\frac{\partial c_{i,w}}{\partial z} + \left(\frac{\varepsilon}{\tau}\frac{D_i^T}{M_i}\frac{1}{T}\right)\frac{\partial T}{\partial z}\right) \tag{6.5}$$

Boundary condition between the resistive heater and the ceramic support (at z = $\delta_{\rm wc}$ + $\delta_{\rm sup}$) is given as

$$\rho_{cr}C_{p,cr}\frac{\partial T_{cr,1}}{\partial t}\Delta z_1^+ = q'' + \lambda_{cr}\frac{\partial T_{cr,1}}{\partial z_1} \tag{6.6}$$

where q'' is the heat flux supplied by the heater. Δz_1^+ is the halfway of the distance between the heater-ceramic support interface and adjacent grid point in the ceramic support. The boundary condition at the ceramic support-washcoat interface (at z = $\delta_{\rm wc}$) is given as

$$\left(\rho_{cr}C_{p,cr}\Delta z_2^+ + \rho_{wc}C_{p,wc}\Delta z_3^+\right)\frac{\partial T_k}{\partial t} = \lambda_{cr}\frac{T_{k-1} - T_k}{\Delta z_2} - \lambda_{eff}\frac{T_k - T_{k+1}}{\Delta z_3} \tag{6.7}$$

where Δz_2 is the distance between the ceramic support-washcoat interface and adjacent grid point in the ceramic substrate. Δz_3 is the distance between the ceramic support-washcoat interface and adjacent grid point in the washcoat. ${\Delta z_2}^+$ and ${\Delta z_3}^+$ are given as ${\Delta z_2}^+ = \Delta z_2/2$ and ${\Delta z_3}^+ = \Delta z_3/2$, respectively. Finally, energy balance at the gas-washcoat interface (at z = 0.0) is given as

$$\left(\rho_{mix}c_{p,mix}\Delta z_4^+ + \rho_{wc}C_{p,wc}\Delta z_5^+\right)\frac{\partial T_{int}}{\partial t} = \lambda\frac{\partial T_{int}}{\partial z_4}$$
$$-\sum_{i=1}^{N_g} h_i(j_i + \rho Y_i u) - \sigma\epsilon\left({T_{int}}^4 - {T_{rad}}^4\right) + \lambda_{eff}\frac{\partial T_{int}}{\partial z_5} \tag{6.8}$$

where Δz_4 is the distance between the gas-washcoat interface and adjacent grid point in the gas-phase. Δz_5 is the distance between the gas-washcoat interface and adjacent grid point in the washcoat. Δz_4^+ and Δz_5^+ are given as $\Delta z_4{}^+ = \Delta z_4/2$ and $\Delta z_5{}^+ = \Delta z_5/2$, respectively. The first term on the right hand side of Eq. (6.8) accounts for heat conduction from the interface to the gas according to the Fourier heat conductivity law. The second term describes convective and diffusive energy transport from the gas-phase to the surface, where h_i is the enthalpy of species *i*. The third term is heat radiation from the surface due to the Stefan-Boltzmann law, where σ is the Stefan-Boltzmann constant and ε is the emissivity of the outer washcoat surface. Here T_{rad} is the reference temperature to which the surface radiates. The fourth term encompasses heat conduction from washcoat to interface according to the Fourier heat conductivity law.
Further, the simulations with the energy balance equations indicate that the temperature gradient inside the washcoat is negligible for CPOX and SR of CH_4. Since the washcoat is a thin layer in general and alumina is a highly conductive material, temperature gradient inside the washcoat is obtained less than 0.5 K for all CPOX cases and less than 0.3 K for all SR cases, respectively.

6.4 Conclusions

CPOX and SR of CH_4 are investigated numerically in stagnation flow over a catalytic disk. The chemical and physical processes inside the washcoat are discussed in detail. Internal mass transfer limitations are accounted by using the η-approach, RD-*approach* and *DGM* surface models. The simulations with all three surface models indicate strong diffusion limitations inside the washcoat for all studied CPOX and SR cases. Therefore, internal mass transfer limitations must be considered for accurately inteprating the experiments in CPOX and SR of CH_4 over a thick catalyst layer.

The *RD-approach* and *DGM* give an insight into the reaction routes inside the washcoat. According to the *RD-approach* and *DGM* simulations, there is not a direct reaction mechanism in the catalyst for CPOX cases. At steady state, total oxidation, steam and dry reforming of CH_4 and WGS reactions occur in the catalyst. However, steam reforming is the only reaction route in SR of CH_4.

DGM simulations gives almost identical species profiles with the *RD-approach* for all CPOX and SR cases, which indicates that the species transport inside the washcoat due to pressure-driven convective flow can be neglected.

The simulations show that increasing the reactor pressure and decreasing the inlet flow velocity increases the external mass transfer limitations and decreases the internal mass transfer limitations. Increasing the reactor pressure and decreasing the inlet flow velocity increases the syngas production significantly.

7 Summary and Outlook

This thesis focused on the development and validation of a numerical model for the stagnation-flow reactor (SFR) configuration over porous catalytic surfaces. Therefore, initially a one-dimensional (1D) mathematical model is developed. The mathematical model is based on the 1D flow assumptions with energy and species continuity equations. The mathematical model was established through a newly developed software tool, DETCHEM^{STAG}. The code and model has the advantage (over the alternative codes) of incorporating different models for internal diffusion in the porous catalyst layer and coupling the model with multi-step heterogeneous reaction mechanisms. In this sense, mass transfer in the washcoat was considered for two different conditions, i.e. instantaneous diffusion (infinitely fast mass transport) at the gas-washcoat interface and finite diffusion within the porous layer. Finite diffusion inside the washcoat was accounted from simple to more detailed transport models, i.e., *η-approach*, *RD-approach* and *DGM*. Since these transport models are frequently used in catalytic reactor simulations, it was important to compare their accuracy at a fundamental level.

Further in the thesis, the numerical model and the computer code were applied to study the direct oxidation of CO over Rh/Al_2O_3 catalysts in a SFR. A recently established SFR [98] was used to provide the experimental data and the physical properties of the catalyst. The numerical model and computer code was validated through the experimental results. The results showed that the internal mass transfer limitations were prominent on the system in CO oxidation. Therefore, simulations with the *∞-approach*, which neglects the internal mass transport effects, were unable to make accurate predictions of the measured species profiles. The overall reaction rate and therefore species profiles were strongly influenced by internal mass transport limitations requiring adequate models. In this case, *η-approach* and *RD-approach* predicted the measured species profiles well. The values of η and Φ for CO oxidation results were provided. Since CO oxidation is one of the most prominent reactions, which has been used in fundamental studies in the surface science and catalysis communities, these results also clearly showed that mass transfer limitations have often to be taken into account when interpreting the overall measured reaction rates. In this sense, CO oxidation does not represent a simple system and surface science studies should be more aware of mass transport effects.

The results of DETCHEM^{STAG} and CHEMKIN SPIN codes were compared based on a so-called effective $F_{cat/geo.}$ It was shown that both codes produced almost the same results for the considered cases. The results of direct and indirect coupling of the gas-phase and washcoat equations are discussed. It was revealed that both methods give the same results, when the steady-state results are concerned. However, it is expected that they will produce different results in transient studies. Since the concentration and temperature gradients change steeply near the outer catalyst surface, the results with equidistant grids deviated from the experiments. Therefore, fine mesh generation was always used in the results. The results were produced with simple aspect ratio, but adaptive gridding results were also provided for certain cases. Adaptive gridding does not require user prediction for grid generatio. Instead, it automatically inserts new grid points based on the considered control algorithms.

Subsequently, WGS and RWGS were investigated numerically in stagnation flow over the 100 µm catalytic disk based on the experimental conditions given in [99]. Simulations with the 1D SFR model predicted the experiments, well. According to *η-approach* and *RD-approach* simulations, internal mass transfer limitations were important on the system in WGS and RWGS cases as well. However, the effect of external mass transfer limitations was negligible. The values of η and Φ for WGS and RWGS results were provided. The effect of convective flow on species transport in the washcoat was investigated with *DGM* simulations. In this respect, *DGM* simulations gave identical species profiles with the *RD-approach* simulations for washcoat applications due to low pressure gradients. This reveals that the effect of convective flow on species transport in the washcoat is negligible in WGS catalysts. The simulation results showed that decreasing the inlet velocity and the washcoat thickness and increasing the reactor pressure results in increasing the CO consumption on the surface. It was shown that internal mass transfer limitations and CO consumption at the catalyst surface are strongly affected by the mean pore diameter, tortuosity and porosity of the catalyst. Accordingly, thinner washcoat layers along with the higher reactor pressures (3 atm), lower inlet flow velocities (0.2-0.3 m/s), higher washcoat mean pore diameter, higher washcoat porosity and lower washcoat tortuosity would result in a high CO consumption in WGS catalysts.

Lastly the code was applied to investigate CPOX and SR of CH_4 in stagnation flow over the catalytic disk based on the experiments of [98]. The results showed that both internal and external mass transfer limitations were important on the system. According to the *RD-approach* and *DGM* simulations, there is not a direct syngas formation in the catalyst for CPOX cases. Total oxidation, steam and dry reforming of CH_4 and WGS reactions occur in the catalyst. However, steam reforming is the only reaction route in SR of CH_4. *DGM* simulations showed that the effect of convective flow in the washcoat in CPOX and SR of CH_4 cases are also negligible. According to the simulations, increasing the reactor pressure and decreasing the inlet flow velocity increases the external mass transfer limitations and decreases the internal mass transfer limitations. The results showed that syngas production in CPOX and SR cases can be increased with the increasing the reactor pressure and decreasing the inlet flow velocities.

Further in the thesis, heat transport effects in the washcoat are investigated based on CPOX and SR of CH_4. Simulations with the energy balance equations in the washcoat show that temperature gradient inside the washcoat is negligible. Therefore, it is a proper assumption to consider the washcoat as isothermal in catalytic reactor simulations. *η-approach* was the simplest surface model in this study. Simulations with *η-approach* predicted the experiments reasonably well for many simulations. Therefore, the model offers an alternative to consider the internal mass transfer limitations in practical reactor simulations. *η-approach* offers a computationally inexpensive solution. However, the model does not enable detailed realization of the spatial profiles of species in the washcoat. In this respect, *RD approach* and *DGM* offer an alternative to investigate the species profiles in the washcoat in detail. Since *DGM* couples fluxes of each species with one another, chemical reactions and mass conservation equations, its computational cost is high. In addition, the effect of convective flow in the washcoat is negligible. Therefore, it would be advantageous to use the *RD-approach* over *DGM* due to computational expenses for detailed investigation in washcoat applications, if pressure gradients in the washcoat is negligible. *DGM* could be advantageous only for modeling the systems, where high pressure gradients are likely to occur (such as SOFC anode or membrane reactors).

The numerical model presented in this thesis offers a viable and reliable alternative to investigate the steps of heterogeneous catalytic processes in catalytic reactors. Different numerical models, from simple to more detailed, such as plug-flow reactor model, boundary-layer approach and three-dimensional (3D) computational fluid dynamics (CFD) with heterogeneous reactions are frequently implemented to investigate the steps of heterogeneous catalytic processes. Simple models offer fast simulation, however, they neglect certain physical effects such as diffusive terms (plug flow model etc.). 3D CFD with heterogeneous reactions offers the most comprehensive results for the representation of catalytic reactors but solution expenses also grow excessively due to complex reaction networks. In this case, the 1D SFR model does not neglect certain physical effects, instead it emerges due to natural vanishing of some terms due to mathematical reduction of three dimensional (3D) Navier-Stokes equations. In addition, the results can be assisted by the experiments on the laboratory-scale SFR, which are not viable in practical reactors to perform. Therefore, the developed model can be used for fundamental research regarding the heterogeneous catalytic processes. This thesis investigated the steady-state results based on the steady-state experiments for different reaction networks. In the future, the numerical model can be extended to investigate transient phenomena in catalytic reactors such as catalytic ignition.

References

[1] J.N. Armor, A history of industrial catalysis, Catalysis Today, 163 (2011) 3-9.

[2] B. Lindström, L. Pettersson, A Brief History of Catalysis, CATTECH, 7 (2003) 130-138.

[3] S.H. Fogler, Catalysis and Catalytic Reactors, in: Elements of Chemical Reaction Engineering, Pearson, New Jersey, U.S., 2006, pp. 645-756.

[4] Energy and GHG Reductions in the Chemical Industry via Catalytic Processes, in, Report from International Energy Agency, Dechema Gesellschaft für Chemische Technik und Biotechnologie E.V. and International Council of Chemical Associations Technology Roadmap, 2013.

[5] O. Deutschmann, Computational Fluid Dynamics Simulation of Catalytic Reactors, in: Handbook of Heterogeneous Catalysis, Wiley-VCH Verlag GmbH & Co. KGaA, 2008.

[6] H. Bockhorn, Simulation chemischer Reaktionen in turbulenten Strömungen, Chemie Ingenieur Technik, 63 (1991) 715-724.

[7] C. Houtman, D.B. Graves, K.F. Jensen, Cvd in Stagnation Point Flow - an Evaluation of the Classical 1d-Treatment, Journal of the Electrochemical Society, 133 (1986) 961-970.

[8] R.J. Kee, M.E. Coltrin, P. Glarborg, Stagnation Flows, in: Chemically Reacting Flow, John Wiley & Sons, Inc., 2005, pp. 249-308.

[9] N.E. McGuire, N.P. Sullivan, O. Deutschmann, H.Y. Zhu, R.J. Kee, Dry reforming of methane in a stagnation-flow reactor using Rh supported on strontium-substituted hexaaluminate, Applied Catalysis A-General, 394 (2011) 257-265.

[10] E. Goos, P. Frank, B. Jungfleisch, R. Suntz, H. Bockhorn, Untersuchungen und Modellierungen zu Rußwachstum und Rußoxidation in C2H2 Gegenstromflammen, VDI- Berichte, VDI- Verlag, 1888 (2005) 319.

[11] M. Marquardt, F. Mauß, B. Jungfleisch, R. Suntz, H. Bockhorn, Re-Initiation of Soot Surface Growth in Premixed Counterflow Flames, in:26th Symposium (International) on Combustion, The Combustion Institute, Pittsburgh, 1996, pp. 2343.

[12] O. Deutschmann, F. Behrendt, J. Warnatz, Modeling and Simulation of Heterogeneous Oxidation of Methane on a Platinum Foil, Catalysis Today, 21 (1994) 461-470.

[13] O. Deutschmann, R. Schmidt, F. Behrendt, J. Warnat, Numerical modeling of catalytic ignition, Symposium (International) on Combustion, 26 (1996) 1747-1754.

[14] H. Ikeda, J. Sato, F.A. Williams, Surface Kinetics for Catalytic Combustion of Hydrogen-Air Mixtures on Platinum at Atmospheric-Pressure in Stagnation Flows, Surface Science, 326 (1995) 11-26.

[15] C. Karakaya, O. Deutschmann, Kinetics of hydrogen oxidation on Rh/Al2O3 catalysts studied in a stagnation-flow reactor, Chemical Engineering Science, 89 (2013) 171-184.

[16] A.B. Mhadeshwar, P. Aghalayam, V. Papavassiliou, D.G. Vlachos, Surface reaction mechanism development for platinum-catalyzed oxidation of methane, Proceedings of the Combustion Institute, 29 (2002) 997-1004.

[17] D.G. Vlachos, Homogeneous-heterogeneous oxidation reactions over platinum and inert surfaces, Chemical Engineering Science, 51 (1996) 2429-2438.

[18] J. Warnatz, Resolution of gas phase and surface combustion chemistry into elementary reactions, Symposium (International) on Combustion, 24 (1992) 553-579.

[19] J. Warnatz, M.D. Allendorf, R.J. Kee, M.E. Coltrin, A model of elementary chemistry and fluid mechanics in the combustion of hydrogen on platinum surfaces, Combustion and Flame, 96 (1994) 393-406.

[20] H. Karadeniz, C. Karakaya, S. Tischer, O. Deutschmann, Numerical modeling of stagnation-flows on porous catalytic surfaces: CO oxidation on Rh/Al2O3, Chemical Engineering Science, 104 (2013) 899-907.

[21] B. Ruf, F. Behrendt, O. Deutschmann, S. Kleditzsch, J. Warnatz, Modeling of chemical vapor deposition of diamond films from acetylene-oxygen flames, Proceedings of the Combustion Institute, 28 (2000) 1455-1461.

[22] B. Ruf, F. Behrendt, O. Deutschmann, J. Warnatz, Simulation of homoepitaxial growth on the diamond (100) surface using detailed reaction mechanisms, Surface Science, 352 (1996) 602-606.

[23] A.G. Salinger, R.P. Pawlowski, J.N. Shadid, B.G. Van Bloemen Waanders, Computational analysis and optimization of a chemical vapor deposition reactor with large-scale computing, Industrial & Engineering Chemistry Research, 43 (2004) 4612-4623.

[24] H. Van Santen, C.R. Kleijn, H.E.A. Van Den Akker, Symmetry breaking in a stagnation-flow CVD reactor, Journal of Crystal Growth, 212 (2000) 311-323.

[25] L.L. Raja, R.J. Kee, R. Serban, L.R. Petzold, Computational algorithm for dynamic optimization of chemical vapor deposition processes in stagnation flow reactors, Journal of the Electrochemical Society, 147 (2000) 2718-2726.

[26] R. J. Kee, F. M. Rupley, J. A. Miller, M. E. Coltrin, J. F. Grcar, E. Meeks, H. K. Moffat, A. E. Lutz, G. Dixon-Lewis, M. D. Smooke, J. Warnatz, G. H. Evans, R. S. Larson, R. E. Mitchell, L. R. Petzold, W. C. Reynolds, M. Caracotsios, W. E. Stewart, P. Glarborg, C. Wang, O. Adigun, in, San Diego, CA, 2000.

[27] R.E. Hayes, S.T. Kolaczkowskib, P.K.C. Li, S. Awdry, Evaluating the effective diffusivity of methane in the washcoat of a honeycomb monolith, Applied Catalysis B-Environmental, 25 (2000) 93-104.

[28] F. Behrendt, O. Deutschmann, B. Ruf, R. Schmidt, J. Warnatz, Simulation of Heterogeneous Reaction Systems, in: J. Wolfrum, H.R. Volpp, R. Rannacher, J. Warnatz (Eds.) Gas Phase Chemical Reaction Systems, Springer Berlin Heidelberg, 1996, pp. 265-278.

[29] J. Warnatz, U. Maas, D.W. Robert, Mathematical Description of Premixed Laminar Flat Flames, in: Combustion, Springer Berlin Heidelberg, 2006, pp. 29-38.

[30] J. Warnatz, U. Maas, D.W. Robert, Transport Phenomena, in: Combustion, Springer Berlin Heidelberg, 2006, pp. 57-72.

[31] Y.A. Cengel, J.M. Cimbala, Fluid Kinematics, in: Fluid Mechanics Fundamentals and Applications, Mc Graw Hill Higher Education, 2006, pp. 121-170.

[32] R.J. Kee, M.E. Coltrin, P. Glarborg, Fluid Kinematics, in: Chemically Reacting Flow, John Wiley & Sons, Inc., 2005, pp. 11-66.

[33] R.B. Bird, W.E. Stewart, E.N. Lightfoot, Appendices, in: Transport Phenomena, John Wiley & Sons, Inc., 2007, pp. 807-871.

[34] R.J. Kee, M.E. Coltrin, P. Glarborg, The Conservation Equations, in: Chemically Reacting Flow, John Wiley & Sons, Inc., 2005, pp. 67-149.

[35] R.B. Bird, W.E. Stewart, E.N. Lightfoot, The Equations of Change for Isothermal Systems, in: Transport Phenomena, John Wiley & Sons, Inc., 2007, pp. 75-113.

[36] Y.A. Cengel, J.M. Cimbala, Mass, Bernoulli, and Energy Equations, in: Fluid Mechanics Fundamentals and Applications Mc Graw Hill Higher Education, 2006, pp. 121-170.

[37] Y.A. Cengel, J.M. Cimbala, Momentum Analysis of Flow Systems, in: Fluid Mechanics Fundamentals and Applications, Mc Graw Hill Higher Education, 2006, pp. 227-268.

[38] F. Amir, Z. Yuwen, J. Howell, Generalized Governing Equations, in: Advanced Heat and Mass Transfer, Global Digital Press, Columbia, USA, 2010, pp. 89-208.

[39] B.R. Bird, W.E. Stewart, E.N. Lightfoot, Diffusivity and the Mechanisms of Mass Transport. , in: Transport Phenomena, John Wiley & Sons, Inc., 2007, pp. 513-542.

[40] R.J. Kee, M.E. Coltrin, P. Glarborg, Molecular Transport, in: Chemically Reacting Flow, John Wiley & Sons, Inc., 2005, pp. 487-539.

[41] R.B. Bird, W.E. Stewart, E.N. Lightfoot, Thermal Conductivity and the Mechanisms of Energy Transport, in: Transport Phenomena, John Wiley & Sons, Inc., 2007, pp. 265-289.

[42] S. Mathur, P.K. Tondon, S.C. Saxena, Thermal conductivity of binary, ternary and quaternary mixtures of rare gases, Molecular Physics, 12 (1967) 569-579.

[43] P. N., W. J., Numerical Methods in Laminar Flame Propagation, Friedr. Vieweg & Sohn, 1982.

[44] R.J. Kee, G. Dixon-Lewis, J. Warnatz, M.E. Coltrin, J.A. Miller, A Fortran Computer Code Package For The Evaluation Of Gas-Phase, Multicomponent Transport Properties in, 1986.

[45] R.B. Bird, W.E. Stewart, E.N. Lightfoot, Other Mechanisms for Mass Transport, in: Transport Phenomena, John Wiley & Sons, Inc., 2007.

[46] R.B. Bird, W.E. Stewart, E.N. Lightfoot, Viscosity and the Mechanisms of Momentum Transport, in: Transport Phenomena, John Wiley & Sons, Inc., 2007, pp. 11-39.

[47] J. Warnatz, U. Maas, D.W. Robert, Thermodynamics of Combustion Processes, in: Combustion, Springer Berlin Heidelberg, 2006, pp. 39-56.

[48] M. Wolf, Simulation der katalytischen Methangaskonversion an Platin, in: Naturwissenschaftlich-Mathematicschen Gesamtfakultät, Ruprecht-Karls-Universität Heidelberg, 2000.

[49] R.J. Kee, M.E. Coltrin, P. Glarborg, Heterogeneous Chemistry, in: Chemically Reacting Flow, John Wiley & Sons, Inc., 2005, pp. 445-486.

[50] K. Christmann, Introduction to Surface Physical Chemistry, Springer, New York, 1991.

[51] K.K. Kenneth, Chemical Kinetics and Reaction Mechanisms, in: Principles of Combustion, John Wiley & Sons, Inc., USA, 2005, pp. 116-284.

[52] L. Kunz, L. Maier, S. Tischer, O. Deutschmann, Modeling the Rate of Heterogeneous Reactions, in: Modeling and Simulation of Heterogeneous Catalytic Reactions, Wiley-VCH Verlag GmbH & Co. KGaA, 2011, pp. 113-148.

[53] J. Warnatz, U. Maas, D.W. Robert, Chemical Kinetics, in: Combustion, Springer Berlin Heidelberg, 2006, pp. 73-90.

[54] M. Menzinger, R. Wolfgang, The Meaning and Use of the Arrhenius Activation Energy, Angewandte Chemie International Edition in English, 8 (1969) 438-444.

[55] R.J. Kee, M.E. Coltrin, P. Glarborg, Mass-Action Kinetics, in: Chemically Reacting Flow, John Wiley & Sons, Inc., 2005, pp. 371-400.

[56] A.P.J. Jansen, W.K. Offermans, Lateral Interactions in O/Pt(111): Density-Functional Theory and Kinetic Monte Carlo, in: O. Gervasi, M. Gavrilova, V. Kumar, A. Laganà, H. Lee, Y. Mun, D. Taniar, C. Tan (Eds.) Computational Science and Its Applications – ICCSA 2005, Springer Berlin Heidelberg, 2005, pp. 1020-1029.

[57] F. Hess, H. Over, Kinetic Monte Carlo simulations of heterogeneously catalyzed oxidation reactions, Catalysis Science & Technology, 4 (2014) 583-598.

[58] M. Frenklach, Simulation of surface reactions, Pure and Applied Chemistry, 70 (1998) 477-484.

[59] O. Deutschmann, Interactions between transport and chemistry in catalytic reactors, in: Fakultät für Chemie Ruprecht-Karls-Universität Heidelberg, 2001, pp. 23-24.

[60] O. Deutschmann, S. Tischer, C. Correa, D. Chatterjee, S. Kleditzsch, V. M. Janardhanan, N. Mladenov, H. D. Minh, H. Karadeniz, M. Hettel, in, www.detchem.com, Karlsruhe, 2014.

[61] A.S. Quiney, G. Germani, Y. Schuurman, Optimization of a water-gas shift reactor over a Pt/ceria/alumina monolith, Journal of Power Sources, 160 (2006) 1163-1169.

[62] R.E. Hayes, P.K. Mok, J. Mmbaga, M. Votsmeier, A fast approximation method for computing effectiveness factors with non-linear kinetics, Chemical Engineering Science, 62 (2007) 2209-2215.

[63] R.E. Hayes, A. Fadic, J. Mmbaga, A. Najafi, CFD modelling of the automotive catalytic converter, Catalysis Today, 188 (2012) 94-105.

[64] M.J. Stutz, D. Poulikakos, Optimum washcoat thickness of a monolith reactor for syngas production by partial oxidation of methane, Chemical Engineering Science, 63 (2008) 1761-1770.

[65] J.M. Zalc, S.C. Reyes, E. Iglesia, The effects of diffusion mechanism and void structure on transport rates and tortuosity factors in complex porous structures, Chemical Engineering Science, 59 (2004) 2947-2960.

[66] S.W. Webb, Gas-phase diffusion in porous media: Evaluation of an advective- dispersive formulation and the dusty-gas model including comparison to data for binary mixtures, in: Other Information: PBD: May 1996, 1996, pp. Medium: ED; Size: 72 p.

[67] H.Y. Zhu, R.J. Kee, V.M. Janardhanan, O. Deutschmann, D.G. Goodwin, Modeling elementary heterogeneous chemistry and electrochemistry in solid-oxide fuel cells, Journal of the Electrochemical Society, 152 (2005) A2427-A2440.

[68] J.W. Veldsink, R.M.J. Vandamme, G.F. Versteeg, W.P.M. Vanswaaij, The Use of the Dusty-Gas Model for the Description of Mass-Transport with Chemical-Reaction in Porous-Media, Chemical Engineering Journal and the Biochemical Engineering Journal, 57 (1995) 115-125.

[69] W. Kong, H.Y. Zhu, Z.Y. Fei, Z.J. Lin, A modified dusty gas model in the form of a Fick's model for the prediction of multicomponent mass transport in a solid oxide fuel cell anode, Journal of Power Sources, 206 (2012) 171-178.

[70] V.M. Janardhanan, O. Deutschmann, CFD analysis of a solid oxide fuel cell with internal reforming: Coupled interactions of transport, heterogeneous catalysis and electrochemical processes, Journal of Power Sources, 162 (2006) 1192-1202.

[71] S.A. Seyed-Reihani, G.S. Jackson, Effectiveness in catalytic washcoats with multi-step mechanisms for catalytic combustion of hydrogen, Chemical Engineering Science, 59 (2004) 5937-5948.

[72] Z. Hashin, S. Shtrikman, A Variational Approach to the Theory of the Effective Magnetic Permeability of Multiphase Materials, Journal of Applied Physics, 33 (1962) 3125-3131.

[73] O. Deutschmann, Modellierung von Reaktionen an Oberflächen und deren Kopplung mit chemisch reagierenden Strömungen, in: Naturwissenschaftlich-Mathematicschen Gesamtfakultät, Ruprecht-Karls-Universität Heidelberg, 1996.

[74] G.H. Evans, R. Greif, Forced flow near a heated rotating disk: a similarity solution, Numerical Heat Transfer, 14 (1988) 373-387.

[75] M.E. Coltrin, R.J. Kee, G.H. Evans, A Mathematical-Model of the Fluid-Mechanics and Gas-Phase Chemistry in a Rotating-Disk Chemical Vapor-Deposition Reactor, Journal of the Electrochemical Society, 136 (1989) 819-829.

[76] R.J. Kee, J.A. Miller, G.H. Evans, G. Dixon-Lewis, A computational model of the structure and extinction of strained, opposed flow, premixed methane-air flames, Symposium (International) on Combustion, 22 (1989) 1479-1494.

[77] F. Behrendt, O. Deutschmann, U. Maas, J. Warnatz, Simulation and Sensitivity Analysis of the Heterogeneous Oxidation of Methane on a Platinum Foil, Journal of Vacuum Science & Technology a-Vacuum Surfaces and Films, 13 (1995) 1373-1377.

[78] V. Alopaeus, H. Laavi, J. Aittamaa, A dynamic model for plug flow reactor state profiles, Computers & Chemical Engineering, 32 (2008) 1494-1506.

[79] A. Sari, A. Safekordi, F.A. Farhadpour, Comparison and validation of plug and boundary layer flow models of monolithic reactors: Catalytic partial oxidation of methane on Rh coated monoliths, International Journal of Chemical Reactor Engineering, 6 (2008).

[80] R.J. Kee, M.E. Coltrin, P. Glarborg, Two-Dimensional Systems, in: Chemically Reacting Flow, John Wiley & Sons, Inc., 2005, pp. 693-737.

[81] L. Hui, Mass transport analysis of a showerhead MOCVD reactor, Journal of Semiconductors, 32 (2011) 033006.

[82] N. Mladenov, J. Koop, S. Tischer, O. Deutschmann, Modeling of transport and chemistry in channel flows of automotive catalytic converters, Chemical Engineering Science, 65 (2010) 812-826.

[83] K. E. Brenan , S. L. Campbell , L. R. Petzold Introduction, in: Numerical Solution of Initial-Value Problems in Differential-Algebraic Equations, SIAM, 1996.

[84] M. Sofroniou , R. Knapp Advanced Numerical Differential Equation Solving in Mathematica, in, Wolfram Mathematica, 2008.

[85] A.C. Hindmarsh, Brown P. N., Grant K. E., Lee S. L., Serban R., Shumaker D. E., Woodward C. S., SUNDIALS: Suite of nonlinear and differential/algebraic equation solvers, ACM Trans. Math. Softw., 31 (2005) pp. 363-396.

[86] P. Deuflhard, E. Hairer, J. Zugck, One-step and extrapolation methods for differential-algebraic systems, Numerische Mathematik, 51 (1987) 501-516.

[87] L.R. Petzold, Description of DASSL: a differential/algebraic system solver, 1982.

[88] L. F. Shampine, M. W. Reichelt , J. A. Kierzenka, Solving Index-1 DAEs in MATLAB and Simulink, SIAM, 41 (1997) 538-552.

[89] J. F. Grcar, The Twopnt Program for Boundary Value Problems, in, Sandia National Laboratories, 1992.

[90] O. Großhans, in: Optimal Control of a Reactive Stagnation Point Flow on a Catalytic Plate, Dissertation (Ph.D. Thesis), Naturwissenschaftlich-Mathematicschen Gesamtfakultät der Ruprecht-Karls-Universität Heidelberg, 2001.

[91] H. Karadeniz, C. Karakaya, S. Tischer, O. Deutschmann, Mass transfer effects in stagnation flows on a porous catalyst: Water-Gas-Shift reaction over Rh/Al2O3, Zeitschrift für Physikalische Chemie, Band 229, Heft 5, Seiten 709-737, ISSN (Online) 2196-7156, ISSN (Print) 0942-9352, DOI: 10.1515/zpch-2014-0622 (2014)

[92] A. Wieckowski, R. E. Savinova, G. C. Vayenas, Effects of Size and Contact Structure of Supported Noble Metal Catalysts in Low-Temperature CO Oxidation, in: Catalysis and Electrocatalysis at Nanoparticle Surfaces, Taylor&Francis e-Library, 2009, pp. 645-666.

[93] S.M. McClure, D.W. Goodman, New insights into catalytic CO oxidation on Pt-group metals at elevated pressures, Chemical Physics Letters, 469 (2009) 1-13.

[94] M.J.P. Hopstaken, J.W. Niemantsverdriet, Structure sensitivity in the CO oxidation on rhodium: Effect of adsorbate coverages on oxidation kinetics on Rh(100) and Rh(111), Journal of Chemical Physics, 113 (2000) 5457-5465.

[95] A. Bourane, D. Bianchi, Oxidation of CO on a Pt/Al2O3 catalyst: from the surface elementary steps to light-off tests - V. Experimental and kinetic model for light-off tests in excess of O-2, Journal of Catalysis, 222 (2004) 499-510.

[96] J.A. Anderson, Co Oxidation on Rh/Al2o3 Catalysts, Journal of the Chemical Society-Faraday Transactions, 87 (1991) 3907-3911.

[97] S. Royer, D. Duprez, Catalytic Oxidation of Carbon Monoxide over Transition Metal Oxides, ChemCatChem, 3 (2011) 24-65.

[98] C. Karakaya, A Novel, Hierarchically Developed Surface Kinetics for Oxidation and Reforming of Methane and Propane over Rh/Al2O3,

Dissertation (Ph.D. Thesis), in: Fakultät für Chemie und Biowissenschaften, Karlsruher Institut für Technologie, 2013.

[99] C. Karakaya, R. Otterstätter, L. Maier, O. Deutschmann, Kinetics of the water-gas shift reaction over Rh/Al2O3 catalysts, Applied Catalysis A: General, 470 (2014) 31-44.

[100] L. Maier, B. Schadel, K.H. Delgado, S. Tischer, O. Deutschmann, Steam Reforming of Methane Over Nickel: Development of a Multi-Step Surface Reaction Mechanism, Topics in Catalysis, 54 (2011) 845-858.

[101] P.C. Thune, J.W. Niemantsverdriet, Surface science models of industrial catalysts, Surface Science, 603 (2009) 1756-1762.

[102] R.M. Vanhardeveld, P.L.J. Gunter, L.J. Vanijzendoorn, W. Wieldraaijer, E.W. Kuipers, J.W. Niemantsverdriet, Deposition of Inorganic Salts from Solution on Flat Substrates by Spin-Coating-Theory, Quantification and Application to Model Catalysts, Applied Surface Science, 84 (1995) 339-346.

[103] C. Karakaya, O. Deutschmann, A simple method for CO chemisorption studies under continuous flow: Adsorption and desorption behavior of Pt/Al2O3 catalysts, Applied Catalysis a-General, 445 (2012) 221-230.

[104] R. Zapf, C. Becker-Willinger, K. Berresheim, H. Bolz, H. Gnaser, V. Hessel, G. Kolb, P. Lob, A.K. Pannwitt, A. Ziogas, Detailed characterization of various porous alumina-based catalyst coatings within microchannels and their testing for methanol steam reforming, Chemical Engineering Research & Design, 81 (2003) 721-729.

[105] Y.A. Cengel, J.M. Cimbala, Introduction and Basic Concepts, in: Fluid Mechanics Fundamentals and Applications, Mc Graw Hill Higher Education, 2006, pp. 121-170.

[106] A. Platon, Y. Wang, Water-Gas Shift Technologies, in: Hydrogen and Syngas Production and Purification Technologies, John Wiley & Sons, Inc., 2009, pp. 311-328.

[107] L.C.S. Kahle, T. Roussière, L. Maier, K. Herrera Delgado, G. Wasserschaff, S.A. Schunk, O. Deutschmann, Methane Dry Reforming at High Temperature and Elevated Pressure: Impact of Gas-Phase Reactions, Industrial & Engineering Chemistry Research, 52 (2013) 11920-11930.

[108] T. Roussière, K.M. Schelkle, S. Titlbach, G. Wasserschaff, A. Milanov, G. Cox, E. Schwab, O. Deutschmann, L. Schulz, A. Jentys, J. Lercher, S.A. Schunk, Structure–Activity Relationships of Nickel–Hexaaluminates in Reforming Reactions Part I: Controlling Nickel Nanoparticle Growth and Phase Formation, ChemCatChem, 6 (2014) 1438-1446.

[109] J. Sun, J. DesJardins, J. Buglass, K. Liu, Noble metal water gas shift catalysis: Kinetics study and reactor design, International Journal of Hydrogen Energy, 30 (2005) 1259-1264.

[110] C.A. Cornaglia, J.F. Munera, E.A. Lombardo, Kinetic Study of a Novel Active and Stable Catalyst for the Water Gas Shift Reaction, Industrial & Engineering Chemistry Research, 50 (2011) 4381-4389.

[111] E. Simsek, M. Karakaya, A.K. Avci, Z.I. Onsan, Oxidative steam reforming of methane to synthesis gas in microchannel reactors, International Journal of Hydrogen Energy, 38 (2013) 870-878.

[112] S.K. Ryi, J.S. Park, D.K. Kim, T.H. Kim, S.H. Kim, Methane steam reforming with a novel catalytic nickel membrane for effective hydrogen production, Journal of Membrane Science, 339 (2009) 189-194.

[113] J.T. Richardson, M. Garrait, J.K. Hung, Carbon dioxide reforming with Rh and Pt-Re catalysts dispersed on ceramic foam supports, Applied Catalysis a-General, 255 (2003) 69-82.

[114] G.G. Olympiou, C.M. Kalamaras, C.D. Zeinalipour-Yazdi, A.M. Efstathiou, Mechanistic aspects of the water-gas shift reaction on alumina-supported noble metal catalysts: In situ DRIFTS and SSITKA-mass spectrometry studies, Catalysis Today, 127 (2007) 304-318.

[115] C. Wheeler, A. Jhalani, E.J. Klein, S. Tummala, L.D. Schmidt, The water-gas-shift reaction at short contact times, Journal of Catalysis, 223 (2004) 191-199.

[116] M. Maestri, K. Reuter, Molecular-level understanding of WGS and reverse WGS reactions on Rh through hierarchical multiscale approach, Chemical Engineering Science, 74 (2012) 296-299.

[117] L.D. Schmidt, Catalytic Reactors and Mass Transfer, in: The Engineering of Chemical Reactions, Oxford University Press, 1998, pp. 268-324.

[118] M.E. Davis, R.J. Davis, Effects of Transport Limitations on Rates of Solid-Catalyzed Reactions, in: Fundamentals of Chemical Reaction Engineering, McGraw-Hill, New York, 2003, pp. 184-239.

[119] K. Raweewan, A. Matthias, W. F. Hoelderich, A Review of Mass Transfer Controlling the Reaction Rate in Heterogeneous Catalytic Systems, in: Mass Transfer-Advanced Aspects, InTech, 2011, pp. 667-684.

[120] A. Zamaniyan, A. Behroozsarand, H. Mehdizadeh, H.A. Ghadirian, Modeling of microreactor for syngas production by catalytic partial oxidation of methane, Journal of Natural Gas Chemistry, 19 (2010) 660-668.

[121] K.K. Kim, J.M. Smith, Diffusion in nickel oxide pellets–effects of sintering and reduction, AIChE Journal, 20 (1974) 670-678.

[122] R. Otterstätter, Untersuchung der Wassergas-Shift-Reaktion über Pt und Rh-Katalysatoren, in: Institut für Technische Chemie und Polymerchemie der Fakultät für Chemie und Biowissenschaften, Karlsruher Institut für Technologie, 2012.

[123] N. Laosiripojana, W. Wiyaratn, W. Kiatkittipong, A. Arpornwichanop, A. Soottitantawat, S. Assabumrungrat, Reviews on Solid Oxide Fuel Cell Technology, 2009.

[124] A. Bitsch-Larsen, R. Horn, L.D. Schmidt, Catalytic partial oxidation of methane on rhodium and platinum: Spatial profiles at elevated pressure, Applied Catalysis a-General, 348 (2008) 165-172.

[125] A.L.Y. Tonkovich, B. Yang, S.T. Perry, S.P. Fitzgerald, Y. Wang, From seconds to milliseconds to microseconds through tailored micro-channel reactor design of a steam methane reformer, Catalysis Today, 120 (2007) 21-29.

[126] U. Izquierdo, V.L. Barrio, J.F. Cambra, J. Requies, M.B. Guemez, P.L. Arias, G. Kolb, R. Zapf, A.M. Gutierrez, J.R. Arraibi, Hydrogen production from methane and natural gas steam reforming in conventional and microreactor reaction systems, International Journal of Hydrogen Energy, 37 (2012) 7026-7033.

[127] D.M. Murphy, A. Manerbino, M. Parker, J. Blasi, R.J. Kee, N.P. Sullivan, Methane steam reforming in a novel ceramic micro-channel reactor, International Journal of Hydrogen Energy, 38 (2013) 8741-8750.

[128] F. Wang, B. Qi, G.Q. Wang, L.J. Li, Methane steam reforming: Kinetics and modeling over coating catalyst in micro-channel reactor, International Journal of Hydrogen Energy, 38 (2013) 5693-5704.

[129] R.A. Patil, A. Patnaik, S. Ganguly, A.V. Patwardhan, Effect of structural, thermal and flow parameters on steam reforming of methane in a catalytic microreactor, Chemical Engineering Research & Design, 89 (2011) 2159-2167.

[130] V.V. Kuznetsov, O.V. Vitovsky, O.A. Gasenko, Methane steam reforming in an annular microchannel with Rh/Al2O3 catalyst, Journal of Engineering Thermophysics, 18 (2009) 187-196.

[131] O. Deutschmann, L.D. Schmidt, Modeling the partial oxidation of methane in a short-contact-time reactor, AIChE Journal, 44 (1998) 2465-2477.

[132] C. Appel, J. Mantzaras, R. Schaeren, R. Bombach, A. Inauen, N. Tylli, M. Wolf, T. Griffin, D. Winkler, R. Carroni, Partial catalytic oxidation of methane to synthesis gas over rhodium: in situ Raman experiments and detailed simulations, Proceedings of the Combustion Institute, 30 (2005) 2509-2517.

[133] Z.W. Wei, S.C. Ming, L.W. Hui, Mechanistic study of partial oxidation of methane to syngas using in situ time-resolved FTIR and microprobe Raman spectroscopies, Chemical Record, 2 (2002) 102-112.

[134] K.H. Hofstad, J.H.B.J. Hoebink, A. Holmen, G.B. Marin, Partial oxidation of methane to synthesis gas over rhodium catalysts, Catalysis Today, 40 (1998) 157-170.

[135] D.A. Hickman, E.A. Haupfear, L.D. Schmidt, Synthesis Gas-Formation by Direct Oxidation of Methane over Rh Monoliths, Catalysis Letters, 17 (1993) 223-237.

[136] O. Deutschmann, R. Schwiedernoch, L.I. Maier, D. Chatterjee, Natural gas conversion in monolithic catalysts: Interaction of chemical reactions and transport phenomena, in: J.J.S. E. Iglesia, T.H. Fleisch (Eds.) Studies in Surface Science and Catalysis, Elsevier, 2001, pp. 251-258.

[137] I. Tavazzi, A. Beretta, G. Groppi, P. Forzatti, Development of a molecular kinetic scheme for methane partial oxidation over a Rh/alpha-Al2O3 catalyst, Journal of Catalysis, 241 (2006) 1-13.

[138] R. Horn, K.A. Williams, N.J. Degenstein, A. Bitsch-Larsen, D.D. Nogare, S.A. Tupy, L.D. Schmidt, Methane catalytic partial oxidation on autothermal Rh and Pt foam catalysts: Oxidation and reforming zones, transport effects, and approach to thermodynamic equilibrium, Journal of Catalysis, 249 (2007) 380-393.

[139] D.D. Nogare, N.J. Degenstein, R. Horn, P. Canu, L.D. Schmidt, Modeling spatially resolved profiles of methane partial oxidation on a Rh foam catalyst with detailed chemistry, Journal of Catalysis, 258 (2008) 131-142.

[140] C. Diehm, O. Deutschmann, Hydrogen production by catalytic partial oxidation of methane over staged Pd/Rh coated monoliths: Spatially resolved concentration and temperature profiles, International Journal of Hydrogen Energy, 39 (2014) 17998-18004.

[141] R. Schwiedernoch, S. Tischer, C. Correa, O. Deutschmann, Experimental and numerical study on the transient behavior of partial oxidation of methane in a catalytic monolith, Chemical Engineering Science, 58 (2003) 633-642.

[142] J.G. Jakobsen, M. Jakobsen, I. Chorkendorff, J. Sehested, Methane Steam Reforming Kinetics for a Rhodium-Based Catalyst, Catalysis Letters, 140 (2010) 90-97.

[143] M. Maestri, D.G. Vlachos, A. Beretta, G. Groppi, E. Tronconi, Steam and dry reforming of methane on Rh: Microkinetic analysis and hierarchy of kinetic models, Journal of Catalysis, 259 (2008) 211-222.

[144] M. Zeppieri, P.L. Villa, N. Verdone, M. Scarsella, P. De Filippis, Kinetic of methane steam reforming reaction over nickel- and rhodium-based catalysts, Applied Catalysis a-General, 387 (2010) 147-154.

[145] S. Walter, S. Malmberg, B. Schmidt, M.A. Liauw, Mass transfer limitations in microchannel reactors, Catalysis Today, 110 (2005) 15-25.

[146] O. Korup, C.F. Goldsmith, G. Weinberg, M. Geske, T. Kandemir, R. Schlogl, R. Horn, Catalytic partial oxidation of methane on platinum investigated by spatial reactor profiles, spatially resolved spectroscopy, and microkinetic modeling, Journal of Catalysis, 297 (2013) 1-16.

Appendicies

Appendix A: Deriving the Momentum Equations through Stress Tensors

The equations used in this section for deriving the Navier-Stokes equations through stress tensors were given in [33, 36].

In section 2.1.1.2, conservation of momentum is given for control volume as

$$\sum \vec{F} = \frac{d}{dt}\int_{CV} \rho \vec{V} dV + \int_{CS} \rho \vec{V}(\vec{V}_{rel} \cdot \vec{n}) dA \quad . \tag{A.1}$$

Total external forces on the left hand side of Eq.A1 consist of body forces (gravity, electromagnetic forces) and surface forces (pressure and viscous forces):

$$\sum F = \sum \vec{F}_{body} + \sum \vec{F}_{surface} \tag{A.2}$$

Total body forces on the *control volume* is given as

$$\sum \vec{F}_{body} = \int_{CV} \rho \vec{g}\, dV = m_{CV}\vec{g} \tag{A.3}$$

Surface forces consist of normal and shear stresses. Normal stresses act always through the normal of the surface. Normal and shear stresses are given conveniently in a stress tensor in the Cartesian coordinate system as

$$T = \begin{pmatrix} \tau_{zz} & \tau_{zr} & \tau_{z\theta} \\ \tau_{rz} & \tau_{rr} & \tau_{r\theta} \\ \tau_{\theta z} & \tau_{\theta r} & \tau_{\theta\theta} \end{pmatrix} \tag{A.4}$$

Diagonal elements of the stress tensor are the normal stresses (pressure stresses) and other elements are shear stresses (viscous stresses). Eq.A1 can be rearranged now as

$$\sum F = \sum \vec{F}_{\text{body}} + \sum \vec{F}_{\text{surface}} = \int_{\text{CV}} \rho\, \vec{\text{g}}\, \text{dV} + \int_{\text{CS}} \text{T}\, \vec{\text{n}}\, \text{d}A \qquad \text{A.5}$$

If Eq.A5 is set into Eq.A1, the following equation is obtained

$$\int_{\text{CV}} \rho\, \vec{\text{g}}\, \text{dV} + \int_{\text{CS}} \text{T}\, \vec{\text{n}}\, \text{d}A = \frac{\text{d}}{\text{d}t} \int_{\text{CV}} \rho\, \vec{\text{V}}\, \text{dV} + \int_{\text{CS}} \rho \vec{\text{V}} (\vec{\text{V}}_{\text{rel}} \cdot \vec{\text{n}}) \text{d}A \quad . \qquad \text{A.6}$$

Following differential form of the momentum conservation equation is obtained by using the divergence theorem as [34, 35]:

$$\frac{\partial}{\partial t}(\rho \vec{\text{V}}) + \nabla \cdot (\rho \vec{\text{V}} \otimes \vec{\text{V}}) = \rho \vec{\text{g}} + \nabla \cdot \text{T} \qquad \text{A.7}$$

In this form, the momentum equation is not practical. It is needed to separate pressure stresses and viscous stresses. In addition, viscous stresses can be given in terms of a strain rate tensor Then, stress tensor can be given in cylindrical coordinates as [34]

$$\text{T} = \begin{pmatrix} -P & 0 & 0 \\ 0 & -P & 0 \\ 0 & 0 & -P \end{pmatrix} + \begin{pmatrix} 2\mu \frac{\partial v_z}{\partial z} + \kappa \nabla \cdot \vec{\text{V}} & \mu \left(\frac{\partial v_r}{\partial z} + \frac{\partial v_z}{\partial r} \right) & \mu \left(\frac{\partial v_\theta}{\partial z} + \frac{1}{r} \frac{\partial v_z}{\partial \theta} \right) \\ \mu \left(\frac{\partial v_r}{\partial z} + \frac{\partial v_z}{\partial r} \right) & 2\mu \frac{\partial v_r}{\partial r} + \kappa \nabla \cdot \vec{\text{V}} & \mu \left(\frac{\partial v_\theta}{\partial r} - \frac{v_\theta}{r} + \frac{1}{r} \frac{\partial v_r}{\partial \theta} \right) \\ \mu \left(\frac{\partial v_\theta}{\partial z} + \frac{1}{r} \frac{\partial v_z}{\partial \theta} \right) & \mu \left(\frac{\partial v_\theta}{\partial r} - \frac{v_\theta}{r} + \frac{1}{r} \frac{\partial v_r}{\partial \theta} \right) & 2\mu \left(\frac{1}{r} \frac{\partial v_\theta}{\partial \theta} + \frac{v_r}{r} \right) + \kappa \nabla \cdot \vec{\text{V}} \end{pmatrix} \qquad \text{A.8}$$

where κ is the bulk viscosity and $\nabla \cdot \vec{\text{V}}$ is the velocity divergence. The term $\nabla \cdot \vec{\text{V}}$ becomes zero in case of an incompressible fluid. Inserting Eq.A8 into Eq.A7 gives the following momentum equations in cylindrical coordinates:

In r-component:

$$\rho\left(\frac{\partial v_r}{\partial t}+v_r\frac{\partial v_r}{\partial r}+\frac{v_\theta}{r}\frac{\partial v_r}{\partial \theta}+v_z\frac{\partial v_r}{\partial z}-\frac{v_\theta^2}{r}\right)=-\frac{\partial p}{\partial r}+\rho \mathrm{g}_r$$
$$+\left[\frac{\partial}{\partial r}\left(2\mu\frac{\partial v_r}{\partial r}+\kappa\nabla\cdot\vec{\mathrm{V}}\right)+\frac{1}{r}\frac{\partial}{\partial \theta}\left(\mu\left(\frac{\partial v_\theta}{\partial r}-\frac{v_\theta}{r}+\frac{1}{r}\frac{\partial v_r}{\partial \theta}\right)\right)\right.$$
$$\left.+\frac{\partial}{\partial z}\left(\mu\left(\frac{\partial v_r}{\partial z}+\frac{\partial v_z}{\partial r}\right)\right)+\frac{2\mu}{r}\left(-\frac{1}{r}\frac{\partial v_\theta}{\partial \theta}-\frac{v_r}{r}+\frac{\partial v_r}{\partial r}\right)\right] \tag{A.9}$$

In $\boldsymbol{\theta}$-component:

$$\rho\left(\frac{\partial v_\theta}{\partial t}+v_r\frac{\partial v_\theta}{\partial r}+\frac{v_\theta}{r}\frac{\partial v_\theta}{\partial \theta}+v_z\frac{\partial v_\theta}{\partial z}+\frac{v_r v_\theta}{r}\right)=-\frac{1}{r}\frac{\partial p}{\partial \theta}+\rho g_\theta$$
$$+\left[\frac{\partial}{\partial r}\left(\mu\left(\frac{\partial v_\theta}{\partial r}-\frac{v_\theta}{r}+\frac{1}{r}\frac{\partial v_r}{\partial \theta}\right)\right)+\frac{1}{r}\frac{\partial}{\partial \theta}\left(2\mu\left(\frac{1}{r}\frac{\partial v_\theta}{\partial \theta}+\frac{v_r}{r}\right)+\kappa\nabla\cdot\vec{\mathrm{V}}\right)\right.$$
$$\left.+\frac{\partial}{\partial z}\left(\mu\left(\frac{\partial v_\theta}{\partial z}+\frac{1}{r}\frac{\partial v_z}{\partial \theta}\right)\right)+\frac{2\mu}{r}\left(\frac{1}{r}\frac{\partial v_r}{\partial \theta}+\frac{\partial v_\theta}{\partial r}-\frac{v_\theta}{r}\right)\right] \tag{A.10}$$

In z-component:

$$\rho\left(\frac{\partial v_z}{\partial t}+v_r\frac{\partial v_z}{\partial r}+v_z\frac{\partial v_z}{\partial z}+\frac{v_\theta}{r}\frac{\partial v_z}{\partial \theta}\right)=-\frac{\partial p}{\partial z}+\rho g_z$$
$$+\left[\frac{1}{r}\frac{\partial}{\partial r}\left(r\mu\left(\frac{\partial v_r}{\partial z}+\frac{\partial v_z}{\partial r}\right)\right)+\frac{1}{r}\frac{\partial}{\partial \theta}\left(\mu\left(\frac{\partial v_\theta}{\partial z}+\frac{1}{r}\frac{\partial v_z}{\partial \theta}\right)\right)+\frac{\partial}{\partial z}\left(2\mu\frac{\partial v_z}{\partial z}+\kappa\nabla\cdot\vec{\mathrm{V}}\right)\right] \tag{A.14}$$

Appendix B: Surface Reaction Mechanisms

Table B.1. Surface reaction mechanism for partial oxidation and reforming of CH_4 (all reactions in the list), blue highlighted reactions indicates the reversible WGS reactions

	Reaction	$A^{\dagger}$(cm, mol,s)	$\beta(-)^{\ddagger}$	Ea(kJ/mol)
R1	H_2 + Rh(s) + Rh(s) ⟶ H(s) + H(s)	3.000×10^{-2b}	stick. coeff.	
R2	O_2 + Rh(s) + Rh(s) ⟶ O(s) + O(s)	1.000×10^{-2b}	stick. coeff.	
R3	H_2O + Rh(s) ⟶ H_2O(s)	1.000×10^{-1b}	stick. coeff.	
R4	CO_2 + Rh(s) ⟶ CO_2(s)	4.800×10^{-2b}	stick. coeff.	
R5	CO + Rh(s) ⟶ CO(s)	4.971×10^{-1b}	stick. coeff.	
R6	CH_4 + Rh(s) ⟶ CH_4(s)	1.300×10^{-2b}	Stick .coeff.	
R7	H(s) + H(s) ⟶ Rh(s) + Rh(s) + H_2	5.574×10^{19}	0.239	59.69
R8	O(s) + O(s) ⟶ Rh(s) + Rh(s) + O_2	5.329×10^{22}	-0.137	387.00
R9	H_2O(s) ⟶ H_2O + Rh(s)	6.858×10^{14}	-0.280	44.99
R10	CO(s) ⟶ CO + Rh(s)	1.300×10^{13}	0.295	$134.07-47\theta_{CO}$
R11	CO_2(s) ⟶ CO_2 + Rh(s)	3.920×10^{11}	0.315	20.51
R12	CH_4 (s) ⟶ CH_4 + Rh(s)	1.523×10^{13}	-0.110	26.02
R13	H(s) + O(s) ⟶ OH(s)+ Rh(s)	8.826×10^{21}	-0.048	73.37
R14	OH(s)+ Rh(s) ⟶ H(s) + O(s)	1.000×10^{21}	0.045	48.04
R15	H(s) + OH(s) ⟶ H_2O(s)+ Rh(s)	1.743×10^{22}	-0.127	41.73
R16	H_2O(s) + Rh(s) ⟶ H(s) + OH(s)	5.408×10^{22}	0.129	98.22
R17	OH(s) + OH(s) ⟶ H_2O(s) + O(s)	5.736×10^{20}	-0.081	121.59
R18	H_2O(s) +O(s) ⟶ OH(s) + OH(s)	1.570×10^{22}	0.081	203.41
R19	CO_2(s) + Rh(s) ⟶ CO(s) + O(s)	5.752×10^{22}	-0.175	106.49
R20	CO(s) + O(s) ⟶ CO_2(s) + Rh(s)	6.183×10^{22}	0.034	129.98
R21	CO(s) + Rh(s) ⟶ C(s) + O(s)	6.390×10^{21}	0.000	174.76
R22	C(s) + O(s) ⟶ CO(s) + Rh(s)	1.173×10^{22}	0.000	92.14
R23	CO(s) + OH(s) ⟶ COOH(s) + Rh(s)	2.922×10^{20}	0.000	55.33
R24	COOH(s) + Rh(s) ⟶ CO(s) + OH(s)	2.738×10^{21}	0.160	48.38
R25	COOH(s) + Rh(s) ⟶ CO_2(s) + H(s)	1.165×10^{19}	0.000	5.61
R26	CO_2(s) + H(s) ⟶ COOH(s) + Rh(s)	1.160×10^{20}	-0.160	14.48
R27	COOH(s) + H(s) ⟶ CO(s) + H_2O(s)	5.999×10^{19}	-0.188	33.55
R28	CO(s) + H_2O(s) ⟶ COOH(s) + H(s)	2.258×10^{19}	0.051	97.08
R29	CO(s) + OH(s) ⟶ CO_2(s) + H(s)	3.070×10^{19}	0.000	82.94
R30	CO_2(s) + H(s) ⟶ CO(s) + OH(s)	2.504×10^{21}	-0.301	84.77
R31	C(s) + OH(s) ⟶ CO(s) + H(s)	4.221×10^{20}	0.078	30.04
R32	CO(s) + H(s) ⟶ C(s) + OH(s)	3.244×10^{21}	-0.078	138.26

R33	$CH_4(s) + Rh(s) \longrightarrow CH_3(s) + H(s)$	4.622 x 10^{21}	0.136	72.26
R34	$CH_3(s) + H(s) \longrightarrow CH_4(s) + Rh(s)$	2.137 x 10^{21}	-0.058	46.77
R35	$CH_3(s) + Rh(s) \longrightarrow CH_2(s) + H(s)$	1.275 x 10^{24}	0.078	107.56
R36	$CH_2(s) + H(s) \longrightarrow CH_3(s) + Rh(s)$	1.073 x 10^{22}	-0.078	39.54
R37	$CH_2(s) + Rh(s) \longrightarrow CH(s) + H(s)$	1.275 x 10^{24}	0.078	115.39
R38	$CH(s) + H(s) \longrightarrow CH_2(s) + Rh(s)$	1.073 x 10^{22}	-0.078	52.61
R39	$CH(s) + Rh(s) \longrightarrow C(s) + H(s)$	1.458 x 10^{20}	0.078	23.09
R40	$C(s) + H(s) \longrightarrow CH(s) + Rh(s)$	1.122 x 10^{23}	-0.078	$170.71\text{-}120\theta_C$
R41	$CH_4(s) + O(s) \longrightarrow CH_3(s) + OH(s)$	3.465 x 10^{23}	0.051	77.71
R42	$CH_3(s) + OH(s) \longrightarrow CH_4(s) + O(s)$	1.815 x 10^{22}	-0.051	26.89
R43	$CH_3(s) + O(s) \longrightarrow CH_2(s) + OH(s)$	4.790 x 10^{24}	0.000	114.52
R44	$CH_2(s) + OH(s) \longrightarrow CH_3(s) + O(s)$	2.858 x 10^{21}	0.000	20.88
R45	$CH_2(s) + O(s) \longrightarrow CH(s) + OH(s)$	4.790 x 10^{24}	0.000	141.79
R46	$CH(s) + OH(s) \longrightarrow CH_2(s) + O(s)$	2.858 x 10^{21}	-0.000	53.41
R47	$CH(s) + O(s) \longrightarrow C(s) + OH(s)$	5.008 x 10^{20}	0.000	26.79
R48	$C(s) + OH(s) \longrightarrow CH(s) + O(s)$	2.733 x 10^{22}	0.000	148.81

The rate coefficients are given in the form of $k=AT^{\beta} \exp(-E_a/RT)$; adsorption kinetics is given in the form of sticking coefficients; the surface site density is $\Gamma=2.72 \times 10^{-9}$ mol cm^{-2}

Appendix C: Additional Figures

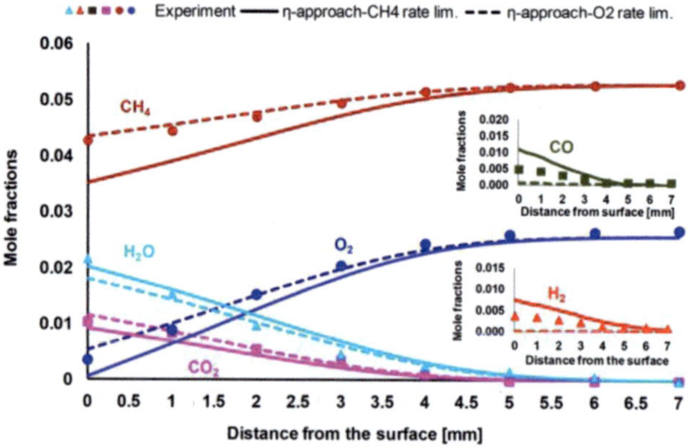

Figure C.1: Comparing *η-approach* simulations for Case 1 (CPOX of CH_4 at 873 K, C/O=1.03, stoichiometric for partial oxidation) by considering CH_4 and O_2 as the rate-limiting species

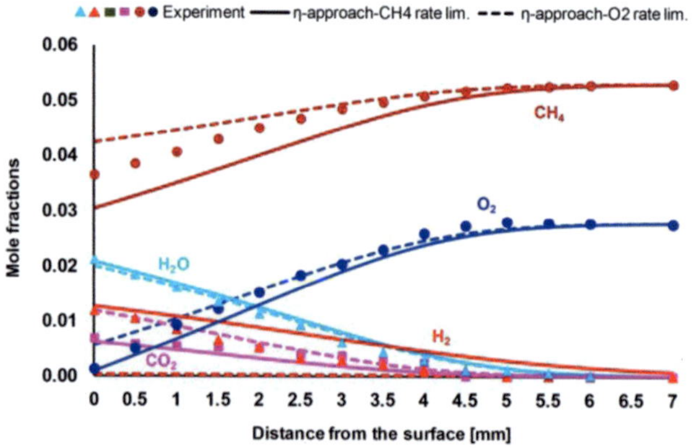

Figure C.2: Comparing *η-approach* simulations for Case 2 (CPOX of CH_4 at 973 K, C/O=0.99, stoichiometric condition for partial oxidation) by considering CH_4 and O_2 as the rate-limiting species

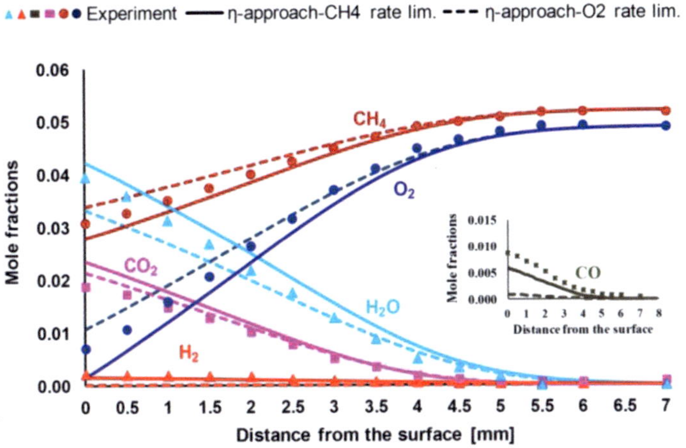

Figure C.3: Comparing *η-approach* simulations for Case 3 (CPOX of CH_4 at 973 K, C/O=0.53, close to the stoichiometric condition for total oxidation) by considering CH_4 and O_2 as the rate-limiting species

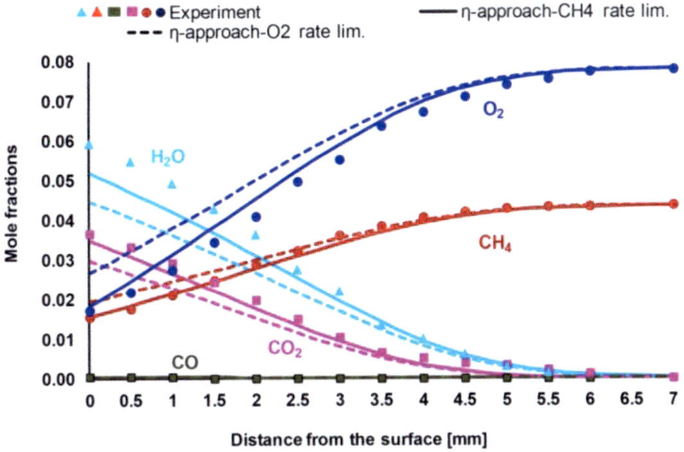

Figure C.4: Comparing *η-approach* simulations for Case 4 (CPOX of CH_4 at 973 K, C/O=0.28, fuel-rich condition for total oxidation) by considering CH_4 and O_2 as the rate-limiting species

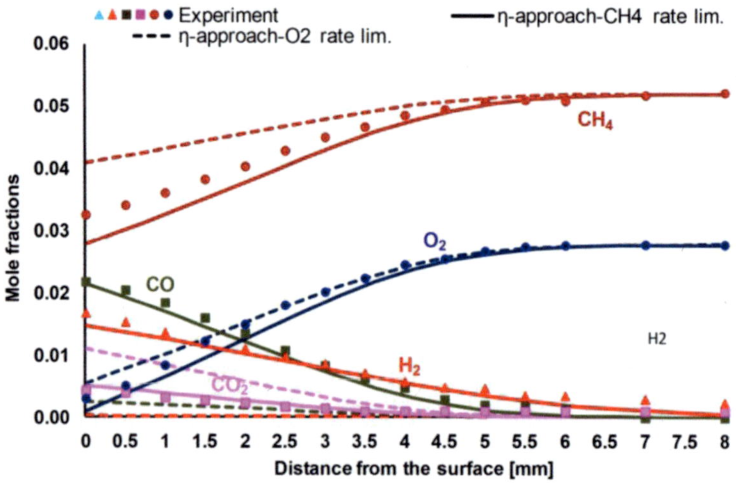

Figure C.5: Comparing *η-approach* simulations for Case 5 (CPOX of CH_4 at 1023 K, C/O=0.93, slightly rich condition (close to the stoichiometry) for the partial oxidation) by considering CH_4 and O_2 as the rate-limiting species

Appendix D: Notation

A	Helmholtz free energy	J/mol
$\Delta_R \bar{A}^o$	standard state Helmholtz free energy	J/mol
A	area	m^2
$A_{\mathrm{geometric}}$	geometric area of the stagnation disc	m^2
A	pre-exponential factor in Arrhenius expression	varies
a_i	Pre-exponential constant in sticking constant expression	
A_{geo}	geometric surface area	m^2
b	intensive variable	
B	extensive variable	
B_{g}	permeability	m^2
b_i	temperature exponent in sticking coefficient expression	
B_{sys}	property B in the system	
B_{cv}	property B in the control volume	
$\dot{B}_{\mathrm{in}}$	influx of property B into the control volume crossing the control surface	
$\dot{B}_{\mathrm{out}}$	outflux of property B from the control volume crossing the control surface	
c_i	concentrations of species i in the washcoat	mol/m^3
$\tilde{c}_i$	activation energy in sticking coefficient expression	J/ mol·K
$c_{i,0}$	species concentrations at the gas-washcoat interface	mol/m^3
$c_{i,\mathrm{b}}$	bulk concentration of species i	mol/m^3
$c_{i,s}$	surface concentration of species i	mol/m^3
$c_{i,\mathrm{w}}$	molar concentration of species i in the washcoat	mol/m^3
c_p	specific heat capacity of the gas mixture	J/kg·K
$c_{p,i}$	specific heat capacity of species i	J/kg·K
$c_{p,s}$	specific heat capacity of the solid support	J/kg·K
$c_{p,\mathrm{wc}}$	specific heat capacity of the gas mixture in the washcoat	J/kg·K
d_{p}	washcoat mean pore diameter	m
d_{pt}	particle diameter	m

d	thickness of the substrate (needed to calculate the conduction losses)	m
D	dispersion	
Da	Damköhler number	
$D_{i,\mathrm{eff}}$	effective diffusion coefficient of species i in the washcoat	m^2/s
D_{ij}	binary diffusion coefficient	m^2/s
$D_{i,\mathrm{Knud}}$	Knudsen diffusion coefficient of species i	m^2/s
$D_{i,\mathrm{M}}$	averaged diffusion coefficient of species i	m^2/s
D_i^{T}	thermal diffusion coefficient of species i	kg/m·s
e_t	total specific energy	J/kg
E_{sys}	total energy of a system	J
E_{a}	activation energy in Arrhenius expression	J/mol
$\vec{\mathrm{F}}$	Force	N
$F_{\mathrm{cat/geo}}$	ratio of the active catalytic surface area to the geometric surface area	
$\vec{\mathrm{g}}$	gravitational acceleration	m/s^2
$\vec{\mathrm{G}}$	general vector quantity	
G	Gibbs free energy	J/mol
$\Delta_R \bar{G}^o$	standard state Gibbs free energy of formation	J/mol
h	Plank's constant	J·s
h	specific enthalpy of the gas mixture	J/kg
h_i	specific enthalpy of species i	J/kg
$h_{i,n}$	elements of the matrix used in the dusty-gas model	
$h_{i,\mathrm{f}}^0$	standard state enthalpy of formation of species i	J/mol
h_{m}	mass transfer coefficient	m/s
$\vec{\mathrm{j}}_i$	corrected diffusive mass flux of species i	$kg/m^2 \cdot s$
$\hat{\mathrm{j}}_i$	diffusive mass flux of species i in the gas-phase	$kg/m^2 \cdot s$
$\vec{\mathrm{j}}_i^{\mathrm{d}}$	diffusive mass flux of species i due to concentration gradient	$kg/m^2 \cdot s$
$\vec{\mathrm{j}}_i^{\mathrm{DGM}}$	diffusive molar flux of species i in the washcoat in *DGM* model	$mol/m^2 \cdot s$
$\mathrm{j}_{i,r}^{\mathrm{DGM}}$	diffusive molar flux of species i in the washcoat in *DGM* model in direction r	$mol/m^2 \cdot s$

$j_{i,\theta}^{\mathrm{DGM}}$	diffusive molar flux of species i in the washcoat in *DGM* model in direction θ	$\mathrm{mol/m^2 \cdot s}$
$j_{i,z}^{\mathrm{DGM}}$	diffusive molar flux of species i in the washcoat in *DGM* model in direction z	$\mathrm{mol/m^2 \cdot s}$
$\vec{j}_i^{\mathrm{T}}$	diffusive mass flux of species i due to temperature gradient	$\mathrm{kg/m^2 \cdot s}$
$j_{i,r}$	diffusive mass flux of species i in direction r	$\mathrm{kg/m^2 \cdot s}$
$j_{i,\theta}$	diffusive mass flux of species i in direction θ	$\mathrm{kg/m^2 \cdot s}$
$j_{i,z}$	diffusive mass flux of species i in direction z	$\mathrm{kg/m^2 \cdot s}$
$\vec{j}_i^{\mathrm{W}}$	diffusive molar flux of species i in the washcoat	$\mathrm{mol/m^2 \cdot s}$
$j_{i,r}^{\mathrm{W}}$	diffusive molar flux of species i in the washcoat in direction r	$\mathrm{mol/m^2 \cdot s}$
$j_{i,\theta}^{\mathrm{W}}$	diffusive molar flux of species i in the washcoat in direction θ	$\mathrm{mol/m^2 \cdot s}$
$j_{i,z}^{\mathrm{W}}$	diffusive molar flux of species i in the washcoat in direction z	$\mathrm{mol/m^2 \cdot s}$
$\vec{j}_q^{\mathrm{c}}$	heat flux due to conduction	$\mathrm{J/m^2 \cdot s}$
$j_{q,r}^{\mathrm{c}}$	heat flux due to conduction in direction r	$\mathrm{J/m^2 \cdot s}$
$j_{q,\theta}^{\mathrm{c}}$	heat flux due to conduction in direction θ	$\mathrm{J/m^2 \cdot s}$
$j_{q,z}^{\mathrm{c}}$	heat flux due to conduction in direction z	$\mathrm{J/m^2 \cdot s}$
$\vec{j}_q^{\mathrm{D}}$	heat flux due to Dufour effect	$\mathrm{J/m^2 \cdot s}$
$\vec{j}_{\mathrm{V}}$	momentum flux	$\mathrm{N/m^3}$
$K_{c,r}$	equilibrium constant in concentration units for reaction r	varies
$K_{p,r}$	equilibrium constant in pressure units for reaction r	
k	rate constant	varies
k_{fr}	rate constant for reaction r in the forward direction	varies
$k_{\mathrm{b},r}$	rate constant for reaction r in the reverse direction	varies
k_{B}	Boltzmann's constant	J/K
L	washcoat thickness	m
m	mass	kg
m_i	mass of species i	kg
m_{ij}	reduced mass of molecules i and j	kg
m_{sys}	mass of the system	kg
$\vec{\mathrm{n}}$	normal unit vector	

M_i	molar mass of species i	kg/mol
$\overline{M}$	mean molar mass of a mixture	kg/mol
N_A	Avagadro's number	1/mol
N_{b}	number of bulk species	
N_{g}	number of gas-phase species	
N_{s}	number of surface species	
p	pressure	Pa
p_{w}	pressure in the washcoat	Pa
p^o	standard pressure at 1 bar	Pa
Q	heat flow into a system	J
$\dot{Q}_{\mathrm{net,in}}$	net heat flow into a system	W
r	radial coordinate	m
R	ideal gas constant	J/ mol·K
s	specific entropy of a mixture	J/kgK
s_i	specific entropy of species i	J/kgK
$s_{i,\mathrm{f}}^0$	standard state entropy of species i	J/kgK
S_i^0	sticking coefficient of species i	
$\mathrm{S}_i^{\mathrm{eff}}$	effective sticking coefficient of species i	
$\dot{s}_{i,\mathrm{eff}}$	effective molar production rate of species i due to surface reactions inside the washcoat	mol/m^2·s
$\dot{s}_{i,\mathrm{w}}$	molar production rate of species i due to surface reactions inside the washcoat (*RD-approach*)	mol/m^2·s
$\dot{s}_i$	molar production rate of species i due to surface reactions	mol/m^2·s
t	time	s
T	temperature	K
T_{b}	backside temperature of the washcoat support	K
T_{rad}	reference temperature in which the surface radiates	K
T_{wc}	temperature of the mixture in the washcoat	K
T^0	temperature at the inlet	K
T_{ij}^*	reduced temperature	
T	stress tensor	N/m^2
$\vec{\mathrm{u}}$	Stefan velocity	m/s
v	fluid velocity	m/s

v_r	cylindrical velocity component in r-direction	m/s
$v_{z,0}$	axial velocity at the inlet	m/s
v_z	cylindrical velocity component in z-direction	m/s
v_θ	cylindrical velocity component in θ-direction	m/s
V	scaled radial velocity	1/s
V	volume	m^3
$\vec{V}$	velocity vector	m/s
$\vec{V}_{rel}$	fluid velocity relative to the control surface	m/s
W	work done on a system	J
$\dot{W}_{net,in}$	net rate of work done on a system	J/s
X_i	mole fraction of species i	
$Y_{i,0}$	mass fraction of species i at the inlet	
Y_i	mass fraction of species i	
z	axial coordinate	m

Greek letters

β	temperature exponent in Arrhenius expression	
ε	catalyst porosity	
ε_{ir}	coverage parameters for species i in reaction r	
ϵ	emissivity of the surface	
η	washcoat effectiveness factor	
λ	thermal conductivity of the mixture	W/m·K
λ_c	thermal conductivity of the washcoat support	W/m·K
λ_{eff}	effective thermal conductivity in the washcoat	W/m·K
λ_i	thermal conductivity of a species	W/m·K
$\lambda_{mix,wc}$	thermal conductivity of the gaseous mixture in the washcoat	W/m·K
λ_{wc}	thermal conductivity of the washcoat substrate	W/m·K
μ	dynamic viscosity of the mixture	kg/m·s
μ_i	dynamic viscosity of the species i	kg/m·s
μ_{ir}	coverage parameters for species i in reaction	
μ_w	dynamic viscosity of the mixture in the washcoat	kg/m·s
ρ_0	density at the inlet	kg/m^3
ρ	density	kg/m^3
ρ_g	density of the gaseous mixture in the washcoat	kg/m^3

ρ_i	density of species i	kg/m^3
σ	collision diameter	m
σ_i	site occupancy number of species i	
Θ_{ij}	length scale in molecular interactions between molecules i and j	m
Θ_i	surface coverage of species i	
θ	angular coordinate	radians
Φ	Thiele modulus	
ϕ	dissipation function	$kg/m^3 \cdot s$
τ	catalyst pore tortuosity	
Λ	eigenvalue of the momentum equation	N/m^4
$\dot{\omega}_i$	molar production rate of species i	$mol/m^3 \cdot s$
$\dot{\omega}_r$	rate-of-progress in reaction r	$mol/m^3 \cdot s$
ζ	chemical potential	J
Ω	volume element	m^3
$\Omega_{ij}^{(1,1)*}$	temperature dependence of the collision integral	
$\vec{\psi}_f^-$	flux through the control surface of the washcoat volume element	
$\vec{\psi}_f^+$	flux through the control surface of the gas-phase volume element	
γ	active catalyst area per washcoat volume	1/m
κ	bulk viscosity	$kg/m \cdot s$
Γ	surface site density	mol/m^2